Earthquake Hazard Assessment

Earthquake Hazard Assessment

India and Adjacent Regions

Sreevalsa Kolathayar & T.G. Sitharam

Department of Civil Engineering, Indian Institute of Science,
Bangalore, India

CRC Press
Taylor & Francis Group
Boca Raton London New York

CRC Press is an imprint of the
Taylor & Francis Group, an **informa** business

A BALKEMA BOOK

Published 2018 by CRCPress/Balkema
P.O. Box 447, 2300 AK Leiden, The Netherlands
e-mail: Pub.NL@taylorandfrancis.com
www.crcpress.com – www.taylorandfrancis.com

First issued in paperback 2021

ISBN 13: 978-0-367-78117-0 (pbk)
ISBN 13: 978-1-138-30923-4 (hbk)

Visit the Taylor & Francis Web site at
http://www.taylorandfrancis.com

and the CRC Press Web site at
http://www.crcpress.com

Typeset by Apex CoVantage, LLC

Library of Congress Cataloging-in-Publication Data
Names: Kolathayar, Sreevalsa, author. | Sitharam, T. G., 1961– author.
Title: Earthquake hazard assessment : India and adjacent regions / Sreevalsa
    Kolathayar & T.G. Sitharam (Department of Civil Engineering, Indian
    Institute of Science, Bangalore, India).
Description: London, UK : CRC Press/Balkema, an imprint of the Taylor &
Francis Group, [2018] | Includes bibliographical references and index.
Identifiers: LCCN 2018011859 (print) | LCCN 2018014709 (ebook) |
    ISBN 9781315143811 (ebook) | ISBN 9781138309234 (hbk)
Subjects: LCSH: Earthquake hazard analysis—India.
Classification: LCC QE537.2.I4 (ebook) | LCC QE537.2.I4 K65 2018 (print) |
    DDC 551.22/0954—dc23
LC record available at https://lccn.loc.gov/2018011859

Contents

Preface   ix
About the authors   xi

I   **Introduction**   I
    1.1   Earthquakes   1
    1.2   Earthquake size   1
          1.2.1   Intensity   1
          1.2.2   Magnitude   2
    1.3   Geological setting of India   3
    1.4   Seismotectonics of Indian subcontinent   5
          1.4.1   Tectonically active shallow crustal region   8
          1.4.2   Subduction zones   8
          1.4.3   Stable continental shield region   9
    1.5   Ground motion prediction equations   9
    1.6   Seismic gap   12
    1.7   Past efforts of seismic hazard studies in India   13

2   **Earthquake data and source models**   17
    2.1   Introduction   17
    2.2   Compilation of the earthquake database   17
    2.3   Homogenization of earthquake magnitude   19
    2.4   Identification of main shocks   22
    2.5   Completeness analysis   25
    2.6   Development of seismotectonic map for India   26
          2.6.1   Scanning of maps   29
          2.6.2   Georeferencing and digitization   30
    2.7   Seismic source models   31
          2.7.1   Linear seismic sources   31
          2.7.2   Point sources   32
          2.7.3   Gridded seismicity source model   33
          2.7.4   Areal sources   34
    2.8   Summary   34

3   Deterministic seismic hazard assessment   **35**
   3.1   Introduction   35
   3.2   Methodology   35
   3.3   Estimation of hazard for Indian subcontinent   36
         3.3.1   Attenuation models   37
         3.3.2   Logic tree structure   40
   3.4   Discussions   40

4   Seismicity analysis and characterization of source zones   **45**
   4.1   Introduction   45
   4.2   Seismicity analysis   45
         4.2.1   Magnitude of completeness   46
         4.2.2   Estimation of a and b values   46
   4.3   Delineation of seismic source zones   51
   4.4   Evaluation of M_{max} for different seismic source zones   52
   4.5   Estimation of seismicity parameters for source zones   54
   4.6   Summary   57

5   Probabilistic seismic hazard assessment   **59**
   5.1   Introduction   59
   5.2   Methodology   59
   5.3   Evaluation of ground motion   61
         5.3.1   Magnitude recurrence rate   61
         5.3.2   Evaluation of hypocentral uncertainty   63
         5.3.3   Uncertainty in attenuation relationship   64
   5.4   PSHA of Indian subcontinent – A case study   64
         5.4.1   Logic tree structure   64
         5.4.2   Evaluation of PGA   65

6   Site response and liquefaction analyses   **77**
   6.1   Introduction   77
   6.2   Site classification   78
   6.3   Site classification methods   78
         6.3.1   Surface geology   78
         6.3.2   Geotechnical data   79
         6.3.3   Geophysical data   80
         6.3.4   Eurocode-8 and NEHRP   81
   6.4   Development of the V_S^{30} map from the topographic slope   83
         6.4.1   Preparation of the slope map   83
         6.4.2   Generation of the V_S^{30} map   83
   6.5   Estimation of surface-level PGA values   86
   6.6   Evaluation of liquefaction potential   91
   6.7   General remarks   95

References   97
Appendix A: Magnitude conversion relations   109
Appendix B: Frequency magnitude distribution plots of seismic source zones   113
Appendix C: Results of PSHA with equal weighting scheme   165
Appendix D: Preparation of the slope map   169
Index   175

Preface

This book presents general principles on earthquakes, data processing, seismicity analysis, seismotectonic maps, ground motion prediction equations, seismic hazard assessments, and liquefaction analysis. The motivation to write this book was to improve the status quo of earthquake hazard studies in technical education in developing countries. In recent years, awareness has increased for the need for earthquake hazard studies to be included in the curriculum of technical courses at the undergraduate and postgraduate levels. This book will impart the necessary knowledge of seismicity and seismic hazards to the professional public through a case study of hazard estimation for Indian subcontinent, a vast region with diverse tectonic characteristics. The reader is expected to gain adequate understanding and information on the following:

1. Homogenization and declustering of the earthquake catalog
2. Ground motion prediction equations (GMPEs)
3. Seismicity analysis for a region
4. Identification and characterization of seismic source zones
5. Source models for seismic hazard analysis
6. Estimation of ground motion parameters through deterministic and probabilistic approaches
7. Evaluation of surface-level seismic hazard, using topographic gradient as a proxy for site amplification
8. Liquefaction susceptibility assessment

These topics are explored through a case study of India and adjacent regions to develop better understanding. The seismicity of the Indian subcontinent is spatiotemporally varied and complex, as the area comprises various tectonic provinces of different seismotectonic and attenuation characteristics.

Thanks to Dr. K. S. Vipin, Specialist Earthquakes, Swiss Re Shared Services, for his assistance in carrying out the hazard assessment. We thank Dr. I. D. Gupta, visiting professor at IIT Roorkee, for his suggestions on the details of source zones. Thanks to Dr. A. Kijko, director of the University of Pretoria Natural Hazard Centre, for sending the updated program for estimating seismicity parameters. We acknowledge the Indian Meteorological Department (IMD); the Indira Gandhi Centre for Atomic Research (IGCAR), Kalpakkam; the National Geophysical Research Institute (NGRI), Hyderabad; the International Seismological Center (ISC); and the United States Geological Survey (USGS) for providing details of earthquake events in the Indian subcontinent and adjoining areas.

We hope graduate students, professors, researchers, scientists, practicing engineers, and policymakers will find this book interesting and valuable. We welcome valuable feedback from the readers for possible improvement of the book. Happy reading!

Sreevalsa Kolathayar

T.G. Sitharam

About the authors

Dr. Sreevalsa Kolathayar pursued M.Tech from IIT Kanpur, Ph.D. from Indian Institute of Science (IISc) and served as International Research Staff at UPC BarcelonaTech Spain with European Union Fellowship. He is presently Asst Professor in the Department of Civil Engineering, Amrita Vishwa Vidyapeetham, Coimbatore, India. Dr. Sreevalsa has authored three books and published several research articles focusing on earthquake hazard assessment and is a reviewer for many international journals. His research interests include Disaster Risk Reduction, Earthquake Preparedness, Geotechnical Earthquake Engg., Seismic Hazard Assessment, Site characterization, Liquefaction Hazard Assessment, Geosynthetics, Water resource management, and Coastal Reservoirs. He served as Chairman of Students Council at IISc and Adviser to Think India, a national forum of students from premier Institutes in India. He is currently the Secretary Indian chapter of International Association for Coastal Reservoir Research (IACRR), Executive Committee Member of Indian Institute of Science Alumni Association and in the board of directors at IIT-Kanpur based start-up G-Intelligence. Dr. Sreevalsa was featured by The New Indian Express in their Edex anniversary edition in 2017 and honored with Edex award: 40 under 40 - South India's Most Inspiring Young Teachers.

Prof. Dr. T.G. Sitharam is a KSIIDC Chair Professor in the area of Energy and Mechanical Sciences and Senior Professor at the Department of Civil Engineering, Indian Institute of Science, Bengaluru (IISc). He was the founder Chairman of the Center for Infrastructure, Sustainable Transport and Urban Planning (CiSTUP) at IISc. He is presently the Chairman of the AICTE South Western Zonal Committee, Regional office at Bengaluru and Vice President of the Indian Society for Earthquake Technology (ISET). He is the founder President of the International Association for Coastal Reservoir Research (IACRR). He was a Visiting Professor at Yamaguchi University, Japan, University of Waterloo, Waterloo, ON, Canada; University of Dalhousie, Halifax, Canada; and ISM Dhanbad, Jharkhand. Prof. Sitharam has obtained his BE(Civil Engg) from Govt BDT College of Engineering from Mysore University, India in 1984, Masters in Civil Engineering from Indian Institute of Science, Bangalore in 1986 and Ph.D. in Civil Engineering from University of Waterloo, Waterloo, Ontario, Canada in 1991 in the area of geomechanics. Further, he was a Research Scientist at center for earth sciences and engineering, University of Texas at Austin, Texas, USA until 1994. Over the last 25 years, he has carried out seismic microzonation of urban centers in India and developed innovative technologies in the area of seismic hazard, liquefaction, fracturing and geotechnical applications, leading to about 500 technical papers, Ten books, four patents, 125 consulting projects and two startup companies. He is the chief editor of two international journals (International Journal of Geotechnical Earthquake Engineering from IGI Global and Journal of Sustainable Urbanization, Planning and Progress (JSUPP) from UDS publishing House, China) in his areas of research. He has guided 30 Ph.D. students and presently he has 8 doctoral students with him. He has advised more than 25 post-doctoral students and 35 M Tech students.

Chapter 1

Introduction

Planet Earth is restless and one cannot control its interior activities and vibrations, like those leading to natural hazards. Earthquakes are one such type of natural hazard that has affected mankind the most. Most of the casualties due to earthquakes happened not because of earthquakes as such, but because of poorly designed structures which could not withstand the earthquake forces. The improper building construction techniques adopted and the high population density are the major causes of the heavy damage due to earthquakes. The damage due to earthquakes can be reduced by following proper construction techniques, taking into consideration appropriate forces on the structure that can be caused due to future earthquakes. The steps towards seismic hazard evaluation are very essential to estimate an optimal and reliable value of possible earthquake ground motion at a region during a specific time period. These predicted values can be an input to assess the seismic vulnerability of an area based on which new construction and the restoration works of existing structures can be carried out.

1.1 EARTHQUAKES

An earthquake is a sudden vibration of the earth caused by an immediate release of energy during rupture of rock that creates seismic waves. The place within the earth's crust where an earthquake originates is called the hypocenter, or focus, of the earthquake (Fig. 1.1). The point vertically above this, on the surface of the earth, is known as the earthquake's epicenter.

1.2 EARTHQUAKE SIZE

Earthquake size can be expressed qualitatively (noninstrumental) or quantitatively (instrumental). It is commonly expressed in terms of intensity or magnitude.

1.2.1 Intensity

Intensity implies how strong an earthquake feels to the observer. It is a qualitative assessment of the damage done by an earthquake. Intensity depends on the distance to the earthquake's epicenter, the strength of the earthquake, and local geology. It is determined by the power of shaking and damage from the quake.

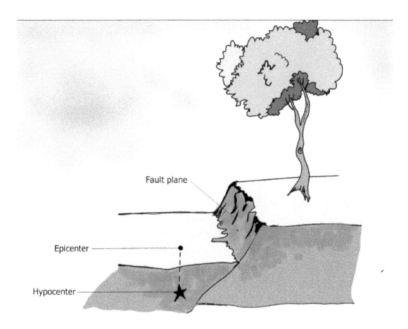

Fault plane

Epicenter

Hypocenter

Figure 1.1 The hypocenter, epicenter, and fault plane

The following are commonly used intensity scales:

- Mercalli-Cancani-Selberg (MCS): 12-level scale used in southern Europe.
- Modified Mercalli (MM): 12-level scale proposed in 1931 by Wood and Neumann. An adapted form is used in North America and some other regions.
- Medvedev-Sponheuer-Karnik (MSK): 12-level scale adopted in Central and Eastern Europe.

Isoseismal intensity scales are used to plot contour lines of equal intensity.

1.2.2   Magnitude

A quantitative measure that is independent of population density and construction type is needed to compare the size of earthquakes worldwide. Magnitude is the quantitative measurement of the amount of energy released by an earthquake.

The Richter magnitude of a local earthquake is a logarithm base 10 of the maximum seismic wave amplitude recorded on a standard seismograph at a distance 100 km from the epicenter. Because earthquake sources are located at all distances from seismographic stations, Richter further developed a method of making allowances for attenuation. Richter scale magnitude is calibrated in such a way that a local magnitude (M_L) of 3 corresponds to an earthquake at a distance of 100 km with a maximum amplitude of 1 mm.

The most common modern magnitude scales are surface wave magnitude and body wave magnitude. Richter's local magnitude does not distinguish between different types of waves.

Surface wave magnitude (M_S) is calculated based on the amplitude of surface (Rayleigh) waves having a period of about 20 s (Gutenberg and Richter, 1956). At large distances from the epicenter, ground motion is dominated by surface waves. Gutenberg and Richter (1936) developed a magnitude scale based on the amplitude of Rayleigh waves:

$$M_S = \log A + 1.66 \log \Delta + 2 \tag{1.1}$$

where A is the maximum ground displacement in micrometers (µm) and Δ is the distance of the seismograph from the epicenter in degrees. This magnitude scale can be used for moderate to large earthquakes having a shallow focal depth, but the seismograph should be 1000 km away from the epicenter (Day, 2002).

Body wave magnitude (m_b) is used to measure deep-focus earthquakes. It is based on the amplitude of the first few cycles of P-wave motion with a period of 1 s (Gutenberg, 1945).

Coda magnitude (M_C) is obtained from characteristics of backscattered waves (Aki, 1969) that follow the passage of primary (unreflected) body and surface waves.

Moment magnitude (M_W) is the most commonly used scale worldwide to determine the magnitude of large earthquakes. The moment magnitude (Kanamori, 1977) can represent the true size of an earthquake because it is based on the seismic moment, which, in turn, is proportional to the product of the rupture area and dislocation of an earthquake fault (Aki, 1966). M_W is defined as:

$$M_W = 2/3 \log_{10} M_0 - 6.05 \tag{1.2}$$

where M_0 is the scalar seismic moment in Newton meter (Nm).

1.3   GEOLOGICAL SETTING OF INDIA

The geological map of India published by the Geological Survey of India is shown in Figure 1.2. This map portrays the standardized stratigraphic classification and correlation of lithostratigraphic units of India. Major rock units are grouped under geological formations indicating different geological ages. The geology of India is so diverse that it contains rocks covering almost the entire range of the geological time scale. It features severely deformed rocks, as well as recently deposited alluvium that has yet to experience diagenesis. Rock type and mineral deposits of many varieties can be found in India. The Indian craton was once part of the supercontinent Pangaea, which later broke apart into two continents: Gondwana and Laurasia. The Indian plate then drifted northward toward the Eurasian plate, leading to the closure of the Tethys Ocean and the creation of the Himalayas and the Tibetan plateau in South Asia. Along with this collision, the Indian plate merged onto the adjacent Australian plate, creating a larger plate, the Indo-Australian plate (Briggs, 2003).

The initial phase of tectonic evolution was marked by the cooling and solidification of the upper layer of Earth's surface in the Archaean era, which is represented by the exposure of gneisses and granites, especially in the peninsular shield. The peninsular shield exhibits geology that has not experienced any significant deformation since the beginning of the Phanerozoic eon. The earliest rocks were severely deformed during the Archaean and Proterozoic eons. They eventually stabilized and formed basements for other rocks to be laid

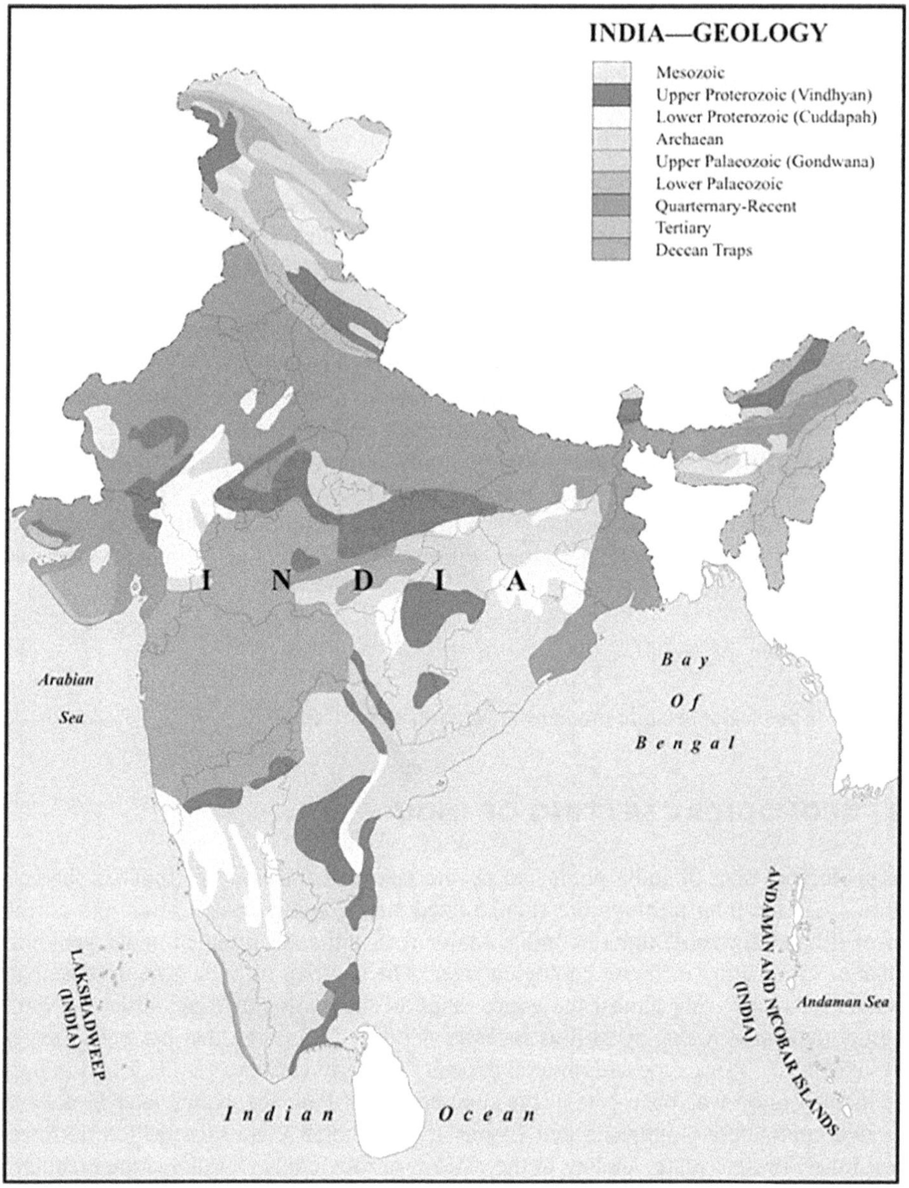
INDIA—GEOLOGY
Mesozoic
Upper Proterozoic (Vindhyan)
Lower Proterozoic (Cuddapah)
Archaean
Upper Palaeozoic (Gondwana)
Lower Palaeozoic
Quarternary-Recent
Tertiary
Deccan Traps
I N D I A
Arabian
Sea
Bay
of
Bengal
LAKSHADWEEP
(INDIA)
ANDAMAN AND NICOBAR ISLANDS
(INDIA)
Andaman Sea
Indian
Ocean

Figure 1.2 Geological map of India

Source: Geological Survey of India.

over them, which are widespread in southern and southeastern India and also in the Aravalli mountain range. The Aravalli range is the remains of an early Proterozoic orogen called the Aravali-Delhi orogen that joined the two older segments that make up the Indian craton; it extends approximately 500 km from its northern end to remote hills and stony ridges into Haryana, ending near Delhi. Minor igneous disturbances, deformation, and successive metamorphism of the Aravalli Mountains represent the primary phase of orogenesis. The erosion of the mountains and added deformation of the sediments of the Dharwaian group was the next phase. The volcanic activities and interruptions associated with this second stage are reflected in the composition of these deposits (Briggs, 2003).

Proterozoic calcareous and arenaceous deposits deposited in the Cuddapah and Vindhyan basins were uplifted during the Cambrian. The sediments are not deformed and have preserved their original horizontal stratification. Early Paleozoic rocks found in the Himalayas include sediments eroded from the crystalline craton and deposited on the Indian platform. In the late Paleozoic, Permo-Carboniferous glaciations left widespread glaciofluvial deposits in central India. These tillites and glacially formed sediments are designated as the Gondwana series. The sediments are overlain by rocks resulting from a Permian marine transgression (Ray, 2006). The Paleozoic of the Indianf peninsula is characterized by river deposits laid down in broad grabens. Deposition by ancient rivers continued until the end of the Mesozoic in various modern rivers such as the Damodar, Godavari, Mahanadi, and Wardha, as well as in the Satpura Hills of Madhya Pradesh.

The late Paleozoic coincided with the deformation and drift of the Gondwana supercontinent, to which the uplift of the Vindhyan sediments and deposition of northern peripheral sediments can be credited. As a result of the drifting of Pangea, large grabens were formed in central India, filling with Upper Jurassic and Lower Cretaceous sandstones and conglomerates. India had separated from Australia and Africa by the Late Cretaceous and was moving northward toward Asia. At the same time, before the Deccan eruptions, uplift in southern India resulted in sedimentation in the adjacent budding Indian Ocean. Exposures of these rocks are seen along the south Indian coast in Pondicherry and Tamil Nadu. One of the most enormous volcanic eruptions in Earth's history occurred at the close of the Mesozoic, resulting in the flowing of Deccan lava. These events mark the final break from Gondwana (Ray, 2006), covering an area of more than 500,000 km^2.

1.4 SEISMOTECTONICS OF INDIAN SUBCONTINENT

The tectonic framework of the Indian subcontinent is spatiotemporally varied and complex. The rapid, high-velocity drifting of the Indian plate toward the Himalayas in the northeastern direction along with its low plate thickness (Kumar et al., 2007) might be the cause of the high seismicity of the Indian region. The Indian plate is moving northwards at about 45 mm per year (mmyr^{-1}), and is colliding with the Eurasian plate (Fig. 1.3, Bilham, 2004). Deformation within Asia reduces India's convergence with Tibet to approximately 18 mm yr^{-1}, as Tibet is extending east to west. This has resulted in the development of potential slip to drive large thrust earthquakes beneath the Himalayas. When continents converge, a great deal of shortening and thickening takes place, as in the Himalayas and Tibet. This massive collision formed the Himalayas, and a great number of earthquakes were generated due to this process. The plate boundary extends from the Himalayan region to the Arakan Yoma and is a major cause of earthquakes in this region. A similar process involving the Indian plate and the Burmese microplate results in earthquakes in the Andaman and Nicobar Islands. The plate boundary areas along the Himalayas and northeast India are characterized by a very high level of seismicity (Gupta, 2006). In addition, earthquakes occur within the Indian shield region, the Indian peninsula, and adjoining parts of the Arabian Sea or the Bay of Bengal. The earthquake zoning map of India divides India into four seismic zones (zones II to V), as shown in Figure 1.4; a previous version consisted of five or six zones. According to the present zoning map, zone V

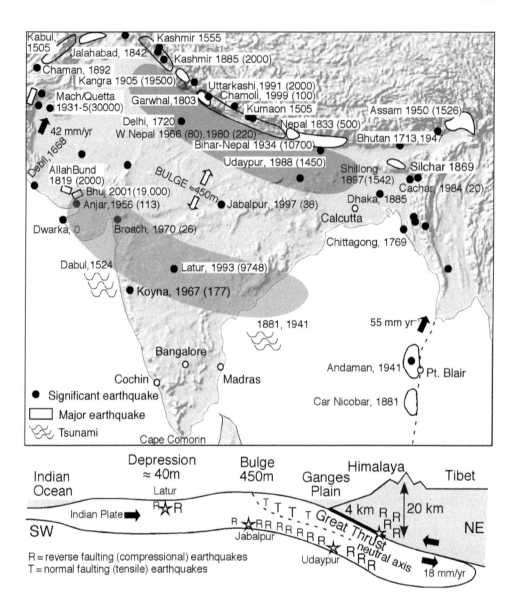

Figure 1.3 Schematic view of Indian tectonics

Source: Bilham (2004).

experiences the highest level of seismicity; zone II is associated with the lowest level of seismicity. Analysis of seismic activity in India can be broadly characterized by three general seismotectonic considerations, as shown in Figure 1.5: tectonically active shallow crustal regions, subduction zones, and stable continental regions (Nath and Thingbaijam, 2010). Subduction zone earthquakes can be further divided into regions with intraslab and interface earthquakes.

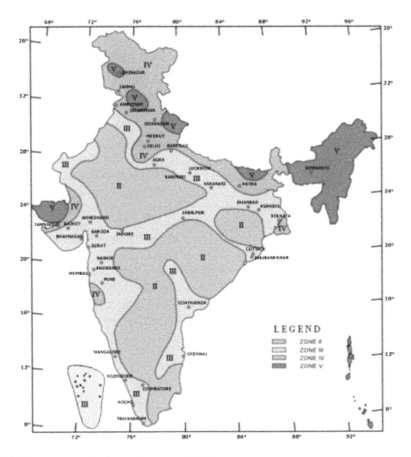

Figure 1.4 Most recent seismic zonation map of India

Source: BIS-1893 (2002).

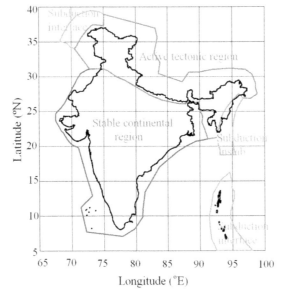

Figure 1.5 Tectonic provinces in and around India

Source: Kolathayar et al. (2012).

1.4.1   Tectonically active shallow crustal region

The seismicity of the Himalayan arc tectonic belt is connected with the underthrusting of the Indian plate beneath the Eurasian plate (Molnar and Tapponnier, 1979; Krishnan, 1953). The tectonically active interplate regions include the Himalayas and southern Tibetan plateau, the northwest frontier province of the Indian plate (Nath and Thingbaijam, 2010; Kayal, 2008). The movement of the Indian plate in the northeast direction and its collision with the Eurasian plate has created the largest mountain range in the world, the Himalayas, with an average height of 4600 m and the most prominent and highest plateau region in the world – the Tibetan plateau. The Indian plate once was one of the fastest moving plates in the world. Before its collision with the Eurasian plate, it attained a very high velocity of around 20 cmyr^{-1} (Kumar et al., 2007). The current movement of Indian plate is estimated to be around 5 cm yr^{-1}. The collision and the subsequent formation of the Himalayas and the Tibetan plateau are associated with very high seismicity.

The entire northeast region falls under zone V of the Indian seismic zonation code (BIS-1893, 2002). This region falls at the junction of the north–south trending Burmese arc and the east–west trending Himalayan arc. Due to this, the entire region has suffered multiple phases of deformation, resulting in the formation of numerous geological structures (Sharma and Malik, 2006).

1.4.2   Subduction zones

The subduction zones include that of Hindukush-Pamir in the northwest frontier province, the Indo-Myanmar arc, and the Andaman-Sumatra seismic belt. Northeast India, especially the region bordering China and Myanmar, is considered the sixth most seismically active region in the world. The seismicity of the Indo-Burmese arc is related to the subduction of the Indian plate underneath the Southeast Asian plate due to the northeastward motion of India (Deshikachar, 1974). The northeastern corner of India, lying between the Himalayan and Burmese arcs, has a complicated seismotectonic setup and unusually high seismic activity (Evans, 1964). The earthquakes in this area are intraslab in nature.

The Andaman and Nicobar Islands are situated on the southeastern side of the Indian landmass. Consisting of about 527 islands, the Andaman and Nicobar Islands are the largest group of islands in the Bay of Bengal. They were formed due to thrust faulting between the Indo-Australian plate and the Burmese plate. These islands are branches of submarine mountains, which are geologically similar to Arakon Yoma of the Myanmar range (Rai and Murty, 2003). The southernmost tip of Great Nicobar Island is only 150 km away from Sumatra, Indonesia. The region is regarded as one of the most seismically active areas in the world. The islands are associated with various geological features, such as subduction tectonics and earthquake processes, crustal deformation, active tectonics, volcanism, etc. The occurrence of the 26 December 2004 earthquake in the southern part of this subduction zone and the tsunami that followed has added strength to the need for comprehensive seismic hazard analysis of the region. The seismotectonic setting of the Andaman and Nicobar Islands has been studied and well documented by various researchers (Eguchi et al., 1979; Mukhopadhyay, 1984, 1988; Dasgupta and Mukhopadhyay, 1993, 1997; Dasgupta et al., 2003; Curray, 2005;Dasgupta et al., 2007a). The Andaman trench, west Andaman fault, Sumatra fault system, Ranong fault, and Khlong Marui fault are active tectonic features in and around the Andaman and Nicobar Islands (Dasgupta et al., 2003, Curray, 2005). The entire island chain lies along the plate boundary between the Indian plate and the Burmese plate. These regions come under subduction zones with interface earthquakes. The Andaman and Nicobar Islands are said to be located on a small tectonic plate, which forms the ridges of the islands, that is packed between the Indo-Australian plate on the west side and the Eurasian plate in the north and the east (Dasgupta et al., 2007b). Spreading centers lie on the eastern side of the islands, and the Indian lithosphere on the western side subducts below the Andaman (Sunda) plate, making this region seismically active (Rajendran and Gupta, 1989). Many damaging earthquakes and tsunami have impacted the Andaman and Nicobar Islands in the past. The Sumatran earthquake of 26 December 2004 occurred along the same source, and the Andaman and Nicobar Islands were one of the worst affected regions during the tsunami. The Andaman and Nicobar Islands have been placed in zone V, the highest level of seismic hazard potential, according to the seismic zonation map of India (BIS-1893, 2002).

1.4.3   Stable continental shield region

Peninsular India is delineated as Stable Continental Region (SCR) with low to moderate seismic activity (Chandra, 1977). The seismicity of this region is intraplate in nature and appears to be associated with some local faults and weak zones (Rao and Murty, 1970). The east-northeast (ENE)–west-southwest (WSW) trending Son-Narmada-Tapti zone is an important tectonic province on the northern edge of the peninsular shield. The major tectonic features in the southern part of the peninsula are the Deccan volcanic province, the southern Indian granulite terrain, the Dharwar craton, the Cuddapah basin, the Godavari and the Mahanadi grabens, and the Eastern and Western Ghats (Gupta, 2006). Researchers, such as Purnachandra Rao (1999), Gangrade and Arora (2000), Reddy (2003),and others, have highlighted the need for seismic study of southern peninsular India. The Bhuj earthquake (26 January 2001; around 19,000 casualties) and Latur earthquake (30 September 1993; around 7928 casualties) were the deadliest earthquakes in this region. Around 10 earthquakes with a magnitude of 6.0 and above have been reported in this region.

1.5   GROUND MOTION PREDICTION EQUATIONS

The ground motion prediction equation (GMPE), or attenuation relation, gives the variation of peak ground acceleration (PGA) at specific structural periods of vibration as a function of earthquake magnitudes and the source-to-site distance, taking the form:

$$\ln y = c_1 + c_2(M - 6) + c_3(M - 6)^2 - \ln R - c_4 R + \ln (\in) \tag{1.3}$$

where y, M, R, and $\in$ refer to PGA/spectral acceleration (g), moment magnitude, hypocentral distance, and the error associated with the regression, respectively.

Table 1.1 lists popular GMPEs developed for different tectonic provinces across the world.

In India, strong motion data are lacking, and this, in turn, has resulted in the development of only very few region-specific GMPEs. Some of the more important GMPEs available in India are Sharma (1998) for the Himalayan region; Iyengar and Ghosh (2004) for the Delhi region; Raghu Kanth and Iyengar (2007) for Peninsular India; Nath et al. (2005) for the Sikkim Himalayas; Nath et al. (2009) for the Guwahati; and Sharma et al. (2009) for the Himalayan region. Of these attenuation relations, the most widely used are those developed

Table 1.1 Recent GMPEs available worldwide

Author (Year)	Region
Iyengar and Ghosh (2004)	Himalayan region
Campbell and Bozorgnia (2003)	Eastern North America
Yuand Wang (2004)	Northeast Tibetan plateau region
Atkinson and Boore (2006)	Eastern North America
Raghu Kanth and Iyengar (2007)	Southern peninsula
Malagnini et al. (2007)	San Francisco
Chiou and Youngs (2008)	Japan, Mexico, California
Castro et al. (2008)	Mexico
Ford et al. (2008)	Northern California
Bennington et al. (2008)	Parkfield, California
Nath et al. (2008)	Garhwal Himalayas
Boore and Atkinson (2008)	Worldwide
Chiou and Youngs (2008)	Japan, Mexico, California
Sharma et al. (2008)	Kachchh region, Gujarat
Chun and Henderson (2009)	North Korea
Sharma et al. (2009)	Himalayan region
Nath et al. (2009)	Guwahati
Ghasemi et al. (2009)	Iran
Koulakov et al. (2010)	Turkey
Gupta (2010)	Indo-Myanmar subduction zone
Graizer (2016)	Central and eastern North America
Zhao et al. (2016)	Japan (subduction slab)
Soghrat and Ziyaeifar (2017)	Northern Iran

by Raghu Kanth and Iyengar (2007) and Sharma et al. (2009). Since only a few attenuation relations were available for the study area, in the present study, we have used some of the more well-accepted GMPEs that have been developed for other regions of the world having similar seismic attenuation characteristics. In a recent study, Nath and Thingbaijam (2010) reviewed ground motion prediction in the Indian scenario with reference to existing GMPEs developed for different tectonic environments and those employed by various regional hazard studies. Figure 1.6 compares some recently developed GMPEs that can be considered for use in Indian regions.

Different sets of GMPEs must be used to model the attenuation properties of the plate boundary region, shield region, and intraslab subduction zones. The relations that can be used for the shield region are those proposed by Campbell and Bozorgnia (2003), Atkinson and Boore (2006), and Raghu Kanth and Iyengar (2007). Of these, the relation by Raghu Kanth and Iyengar (2007) was developed for peninsular Indian shield regions. Raghu Kanth and Iyengar (2007) observed that their model generates predictions similar to those of the available models for other intraplate regions. Attenuation relations given by Campbell and Bozorgnia (2003) and Atkinson and Boore (2006) were developed for eastern

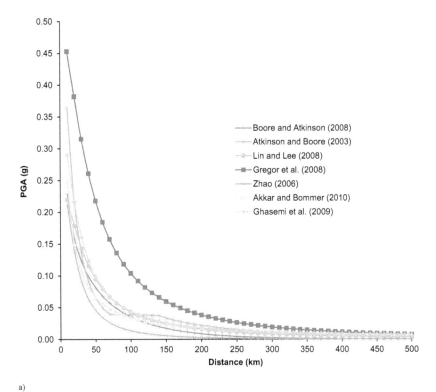

a)

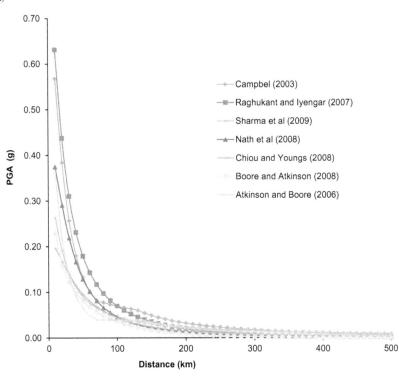

b)

Figure 1.6 Comparison of a few recent and widely recognized attenuation relations

North America (ENA). Based on a study of aftershocks of the Bhuj earthquake, Cramer and Kumar (2003) concluded that the ground motion attenuation in ENA and the peninsular Indian shield are comparable. The similarity of the regional tectonics of ENA and peninsular India has also been noted by Bodin et al. (2004). The GMPEs suitable for active tectonic regions are those put forth by Boore and Atkinson (2008), Sharma et al. (2009), and Akkar and Bommer (2010). Of these, the relation suggested by Sharma et al. (2009) was developed for Himalayan regions of India. Sharma et al. (2009) used data from the Himalayan and Zagros regions, given the considerable similarity of the seismotectonic characteristics of both regions (Ni and Barazangi, 1986). In contrast, the relation by Boore and Atkinson (2008) was developed for active tectonic regions across the world, and the one offered by Akkar and Bommer (2010) was developed for the active tectonic regions of Europe and the Middle East.

Subduction zone earthquakes have different attenuation characteristics compared to shallow crustal earthquakes (Abrahamson and Silva, 1997). Youngs et al. (1997) developed separate attenuation relations for ground motion due to intraslab and interface earthquakes in subduction zone using a global database of about 350 horizontal acceleration components. Atkinson and Boore (2003) updated these relationships using a much bigger worldwide database of about 1200 horizontal acceleration components. Gupta (2010) analyzed a limited number of strong-motion data recorded in the northeast Indian region and concluded that the intraslab earthquakes along the Indo-Burmese subduction zone are found to be characterized by much larger ground motion amplitudes than those for earthquakes along other subduction zones around the world. Gupta (2010) modified attenuation relations developed by Atkinson and Boore (2003) using a global database for subduction zone earthquakes so that they would be more appropriate for northeast India. Attenuation relations suggested by Gupta (2010), Zhao et al. (2006), and Lin and Lee (2008) are capable of predicting ground motion from intraslab subduction earthquakes. For the subduction zone with interface earthquakes, the suitable GMPEs are those from Gregor et al. (2002), Atkinson and Boore (2003), and Lin and Lee (2008). The GMPE by Gupta (2010) is developed specifically for the Indo-Myanmar subduction zone, whereas those offered by Zhao et al. (2006) and Lin and Lee (2008) were developed for the subduction regions (both intraslab and interfaces) of Japan and Taiwan, respectively. Attenuation relations given by Gregor et al. (2002) and Atkinson and Boore (2003) were developed for the Cascadia subduction zone.

1.6   SEISMIC GAP

The *Encyclopedia of Natural Hazards* (Cassidy, 2013) defines a seismic gap as a segment of an active plate boundary that, relative to rest of the boundary, has not recently ruptured and is considered to be more likely to produce an earthquake in the future.

The seismic gap theory (McCann et al., 1979) states that a section of a plate boundary that has not ruptured recently has the greatest chance of rupturing in the future compared to other segments that have experienced large earthquakes. This is based on the understanding that tectonic plates move relative to one another at an approximately constant speed and the assumption that the slip of plate boundary faults occurs primarily during major earthquakes.

A seismic gap is a segment of a fault that has created earthquakes in the past but is at present silent. These are regions along an active fault where stress is accumulating due to

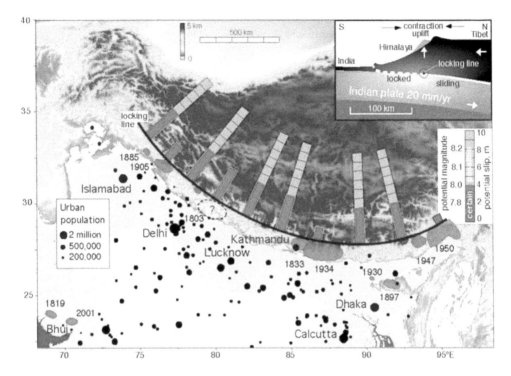

Figure 1.7 Forecast of future earthquakes in seismic gaps of the Himalayas

Source: Bilham et al. (2001).

a lack of recent earthquakes. These regions are considered to be high-risk areas for earthquakes in the near future. Seismic gaps that have developed in the past 500 years along two-thirds of the Himalayas suggest that one or more large earthquakes may be overdue (Bilham, 2004). Figure 1.7 shows the forecast of future earthquakes in seismic gaps of the Himalayas.

1.7   PAST EFFORTS OF SEISMIC HAZARD STUDIES IN INDIA

Various researchers have attempted to evaluate the expected ground motion due to future earthquakes in and around India. Table 1.2 lists important seismic hazard assessments conducted by several researchers for various provinces in India.

A deterministic seismic hazard map of all of India was prepared by Parvez et al. (2003) (Fig. 1.8). This was the first attempt to evaluate the seismic hazard of the Indian subcontinent based on deterministic techniques. The study considered 40 seismogenic sources in India, and regions were classified based on seismicity, tectonics, and geodynamics (Parvez et al., 2003). PGA values were reported for four locations in south India; the maximum PGA value reported was 0.08 g.

Table 1.2 Seismic hazard assessment studies for regions within India

	Study	Study Region
1	Khattri et al. (1984)	Seismic hazard map of India and adjacent areas
2	Bhatia et al. (1999)	PSHA of India and adjoining regions
3	Seeber et al. (1999)	PSHA for Maharashtra
4	BIS-1893 (2002)	Seismic zonation of India
5	Parvez et al. (2003)	DSHA of India
6	Iyengar and Ghosh (2004)	Microzonation of Delhi
7	Raghu Kanth and Iyengar (2006)	PSHA for Mumbai city
8	Nath (2006)	Microzonation of Sikkim Himalayas
9	Sitharam and Anbazhagan (2007)	PSHA for the Bangalore region
10	Jaiswal and Sinha (2007)	PSHA for peninsular India
11	Boominathan et al. (2008)	Seismic hazard assessment of Chennai city
12	Joshi et al. (2007)	DSHA of northeast India
13	Mahajan et al. (2009)	PSHA of northwest Himalayas and adjoining area
14	Vipin et al. (2009)	PSHA of South India
15	Anbazhagan et al. (2009)	PSHA for Bangalore
16	Menon et al. (2010)	PSHA of Tamil Nadu
17	Kolathayar et al. (2012)	DSHA of India
18	Kolathayar and Sitharam (2012)	PSHA of Andaman Nicobar Islands
19	Pallav et al. (2012)	PSHA of Manipur
20	Sitharam and Kolathayar (2013)	PSHA of India using areal sources
21	Vipin et al. (2013)	PSHA of Gujarat
22	Kumar et al. (2013)	PSHA of Lucknow
23	Sil et al. (2013)	PSHA of Tripura and Mizoram
24	Anbazhagan et al. (2014)	PSHA of Coimbatore
25	Desai and Choudhury (2014)	PSHA of Mumbai
26	Nath et al. (2014)	Seismic microzonation of Kolkata
27	Sitharam et al. (2015)	Surface-level PSHA for India
28	Mukhopadhyay and Dasgupta (2015)	PSHA of Kashmir
29	Naik and Choudhury (2015)	DSHA of Goa
30	Nayak et al. (2015)	PSHA of Uttarakhand
31	Anbazhagan et al. (2015)	PSHA of Patna
32	Muthuganeisan and Raghu Kanth (2016)	PSHA of Himachal Pradesh
33	Puri and Jain (2016)	DSHA of Haryana
34	Das et al. (2016)	PSHA of northeast India
35	Anbazhagan et al. (2017)	DSHA and PSHA of Kanpur
36	Dhar et al. (2017)	Seismic Hazard of Odisha by GIS technique
37	Maiti et al. (2017)	PSHA of West Bengal

DSHA, deterministic seismic hazard analysis; PSHA, probabilistic seismic hazard analysis.

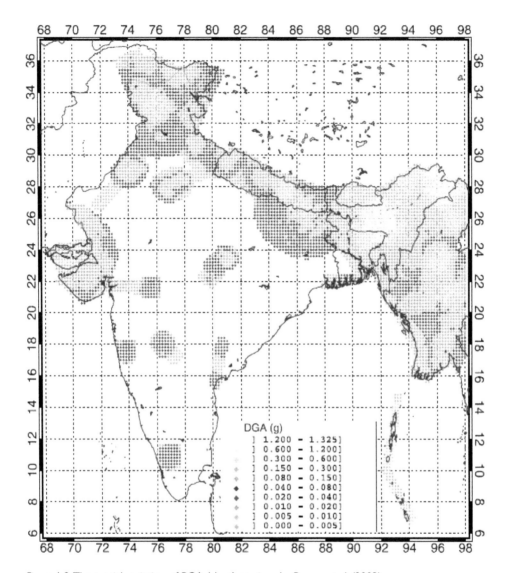

Figure 1.8 The spatial variation of PGA (g) values given by Parvez et al. (2003)

Khattri et al. (1984) developed a PGA hazard map with 10% annual probability of exceedance in 50 years with the use of the attenuation relation developed by Algermissen and Perkins (1976). A PGA hazard map for all of India with 10% probability of exceedance in 50 years was presented by Bhatia et al. (1999) using the attenuation relation of Joyner and Boore (1981) as a part of Global Seismic Hazard Assessment Program (GSHAP), which is shown in Figure 1.9. However, the results obtained byKhattri et al. (1984) and Bhatia et al. (1999) are highly debatable because they used a single attenuation relation for the entire

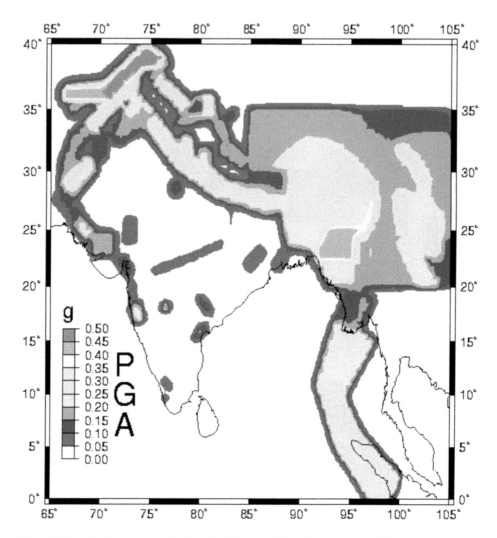

Figure 1.9 Seismic hazard map of India with 10% probability of exceedance in 50 years

Source: Bhatia et al. (1999).

country. Iyengar et al. (2010) developed a probabilistic seismic hazard map for the Indian landmass using linear seismic sources under the aegis of the Indian government's National Disaster Management Authority. Several other efforts have been made by researchers to estimate the seismic hazard for various isolated regions and provinces in the country using different methodologies, as listed in Table 1.2.

Chapter 2

Earthquake data and source models

2.1 INTRODUCTION

A complete and consistent catalog of earthquakes in a region can offer useful data for studying the distribution of earthquakes in an area with respect to space, time, and magnitude. However, most catalogs do not report earthquake magnitudes consistently over time. This raises difficulties for defining seismicity patterns or for assessing seismic hazards. As earthquake magnitude has become a crucial source parameter of earthquakes since its commencement (Richter, 1935), it is essential to convert the original magnitudes based on various scales in different time periods to a standard magnitude scale throughout the whole period. The magnitude scales used in the catalogs of earthquakes in India are not homogeneous. Development of a complete catalog of earthquakes with uniform magnitudes is essential for defining seismicity patterns and assessing seismic hazards for a region. Hence, the original magnitudes of earthquakes have to be converted to a universal and reliable magnitude scale using appropriate magnitude correlations. In the raw catalog, many events can be dependent events that occurred in association with the main shock in a cluster. These aftershocks and foreshocks have to be removed from the catalog using a declustering algorithm to ensure a Poisson distribution of earthquake events. This chapter presents the methodologies employed in processing the earthquake catalog to achieve the above-mentioned objectives. Development of a seismotectonic map for India and various source models suitable for hazard analysis also are described in detail.

2.2 COMPILATION OF THE EARTHQUAKE DATABASE

Extensive data sets from the literature and those provided by numerous national and international agencies were used to create a complete earthquake catalog for India and adjoining areas. Two types of earthquake catalogs were compiled: historical and instrumental. The historical part of the catalog was gathered from the literature. Details of earthquake events for the period from 250BC to 1505AD were obtained from Dunbar et al. (1992). The catalog developed by Dunbar et al. (1992) lists historical earthquakes occurring worldwide from 250 BC to 1991 AD. The later portion of historical earthquakes was compiled from the work of various researchers (Oldham, 1883; Basu, 1964; Kelkar, 1968; Tandon and Srivastava, 1974; Rastogi, 1974; Chandra, 1977, 1978; Kaila and Sarkar, 1978; Rao and Rao, 1984; Srivastava and Ramachandran, 1985; Biswas and Dasgupta, 1986; Guha and Basu, 1993; Bilham, 2004; etc.). A major portion of the instrumental catalog was compiled from several national and international agencies. National agencies included the Guaribidanur Array (GBA), the Indian Meteorological Department (IMD), the Indira Gandhi Centre for Atomic Research (IGCAR), and the National Geophysical Research Institute (NGRI). International agencies included the International Seismological Center (ISC) data file (for the period between 1964 and 2010), and the U.S. Geological Survey NEIC catalog (for the period between 1973 and 2010). Earthquakes that occur outside the study area will also add to the seismic hazard of the study area (US Nuclear Regulatory Commission, 1997). Hence, details of past earthquakes and seismic sources were collected from an area (seismic study area) that extended up to 500 km from the boundary of India.

A catalog of 272,156 earthquakes with magnitudes between M_W 1.0 and M_W 9.0 since 250BC is the basis of the present study. Because the data were collected from various sources, many of the major events were repeated in the catalog, as they were reported by more than one agency/literature. Hence, duplicate events were identified and removed by comparing the location, time, and magnitude of each event. A total of 68,678 events were found to be duplicates, and hence the catalog was refined by considering only the 203,448 original events. The spatial distribution of the epicenters of these events is presented in Figure 2.1. It is clear that the great majority of earthquakes are spread over the Andaman and Nicobar Islands region and the area of north and northeast India that adjoins the Himalayas.

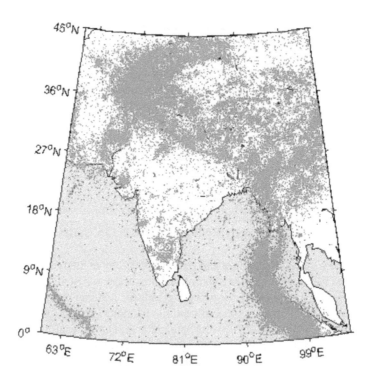

Figure 2.1 Spatial distribution of the epicenters of earthquake events, including aftershocks and foreshocks, for the period from 250BC to 2010AD

Source: Kolathayar et al. (2012).

2.3   HOMOGENIZATION OF EARTHQUAKE MAGNITUDES

A complete earthquake catalog must have a uniform magnitude scale for denoting the size of earthquakes so that a reliable parameterization of the magnitude distribution that is homogeneous and complete with respect to time and size is used in the hazard analysis. The earthquake data obtained were in different magnitude scales, such as body wave magnitude (m_b), surface wave magnitude (M_S), local magnitude (M_L), moment magnitude (M_W), and the earthquake intensity scale (I).

Unfortunately, many of the magnitude scales are limited by saturation toward large earthquakes with $m_b > 6.0$, $M_L > 6.5$, and $M_S > 8.0$. The existence of different magnitude scales necessitates the conversion of these magnitude scales to a single magnitude scale for analysis purposes. The moment magnitude (Kanamori, 1977) can represent the true size of earthquakes because it is based on the seismic moment, which, in turn, is proportional to the product of the rupture area and dislocation of an earthquake fault (Aki, 1966). M_W is defined as:

$$M_W = 2/3 \log_{10} M_0 - 6.05 \tag{2.1}$$

where M_0 is the scalar seismic moment in Nm. The homogenization of the earthquake catalog involves expressing the earthquake magnitudes in one common scale. Practical problems, such as seismic hazard assessment, necessitate the use of a homogenized catalog. Because M_W does not saturate, it is the most reliable magnitude scale for describing the size of an earthquake (Scordilis, 2006). Given that the moment magnitude scale is the most advanced and widely used magnitude scale, the original magnitudes of Indian earthquakes in different time periods have been converted to unified M_W magnitudes. Several relations were proposed by different researchers to convert different magnitude scales to M_W (Nuttli, 1983; Giardini, 1984; Kiratzi et al., 1985; Heaton et al., 1986; Patton and Walter, 1993; Johnston, 1996; Papazachos et al., 2002; Scordilis, 2006; Thingbaijam et al., 2008; among many others). In this study, two methods for magnitude conversion were used; one based on Scordilis (2006) and the other using the developed correlations from the data available for the study area (as detailed below).

Based on the earthquake data available from Indian subcontinent, Kolathayar et al. (2012) developed linear relations connecting various magnitude scales with the moment magnitude scale. Using the available data, an attempt also was made to fit the trend line using polynomial relations. Not much improvement was seen in the regression compared to the linear relation. Of the data sets, 1850 had M_W and m_b, 69 had M_W and M_L, and 1254 had M_W and M_S. A relation connecting M_S with m_b was also developed using 16,734 data sets available in the raw catalog.

The correlation of M_W and m_b (Fig. 2.2) is seen to follow the relation:

$$M_W = 1.08(\pm 0.0152)\, m_b - 0.325(\pm 0.081) \qquad 4 \leq m_b \leq 7.2 \qquad R^2 = 0.732 \tag{2.2}$$

The distribution of M_W versus M_L (Fig. 2.3) gives the correlation as:

$$M_W = 0.815(\pm 0.04)\, ML + 0.767(\pm 0.174) \qquad 3.3 \leq M_L \leq 7 \qquad R^2 = 0.884 \tag{2.3}$$

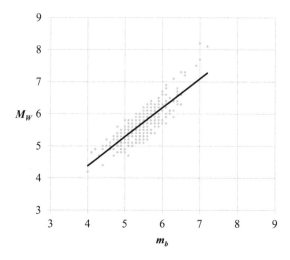

Figure 2.2 Relation between m_b and M_W, $R^2 = 0.719$, $n = 1850$

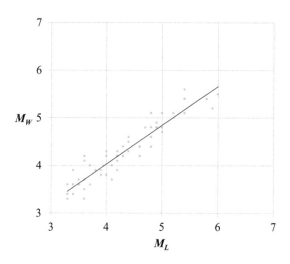

Figure 2.3 Relation between M_L and M_W, $R^2 = 0.884$, $n = 69$

It is found that the M_W obtained from the above relation is very close to the value of observed M_L, which justifies the normal practice of considering M_L as equal to M_W in the seismicity analysis.

The regression between M_W and M_S (Fig. 2.4) yields the following relation:

$$M_W = 0.693(\pm 0.006) \, MS + 1.922(\pm 0.035) \qquad 3.7 \leq M_S \leq 8.8 \qquad R^2 = 0.90 \qquad (2.4)$$

Similarly, the correlation between m_b and M_S (Fig. 2.5) yields the following relation:

$$M_S = 1.057(\pm 0.006) \, mb - 0.649(\pm 0.028) \qquad 3.4 \leq m_b \leq 7 \qquad R^2 = 0.659 \qquad (2.5)$$

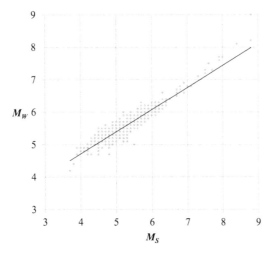

Figure 2.4 Relation between M_S and M_W, $R^2 = 0.893$, $n = 1254$

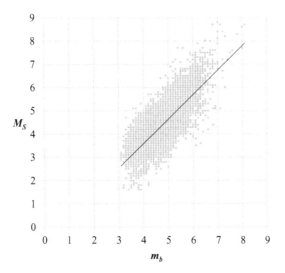

Figure 2.5 Relation between m_b and M_S, $R^2 = 0.659$, $n = 16,734$

The converted M_W magnitudes obtained from Scordilis (2006) and Kolathayar et al. (2012) are in good agreement. The comparison of M_W obtained from m_b and M_S using different relations is shown in Tables 2.1 and 2.2. The relations between different magnitude scales depend on observation errors and source characters such as stress drop, fault geometry, etc. (Heaton et al., 1986). It is always advisable to use the region-specific magnitude conversion relations (Liu et al., 2007). Hence, the M_W values obtained from the correlations developed

Table 2.1 Comparison of magnitude conversion relations relating m_b with M_W

m_b	M_W		
	Kolathayar et al. (2012)	*Scordilis (2006)*	*Thingbaijam et al. (2008)*
3.5	3.9	4.0	3.0
4.0	4.4	4.4	3.7
4.5	4.8	4.9	4.4
5.0	5.3	5.3	5.1
5.5	5.7	5.7	5.8
6.0	6.2	6.1	6.4

Table 2.2 Comparison of magnitude conversion relations relating M_S with M_W

M_S	M_W		
	Kolathayar et al. (2012)	*Scordilis (2006)*	*Thingbaijam et al. (2008)*
3.0	4.0	4.1	3.9
4.0	4.7	4.8	4.6
5.0	5.4	5.4	5.3
6.0	6.1	6.1	6.0
7.0	6.8	7.0	6.7

for the study area were used for homogenization of different magnitude scales. For the conversion of intensity scale to M_W, the relation developed by Menon et al. (2010) was used. Thus, a consistent catalog with a unified magnitude scale was obtained for the entire study area (0° to 45°N and 60° to 105°E). Appendix A presents the magnitude conversion relations developed using the data from active tectonic regions and stable continental regions separately.

2.4   IDENTIFICATION OF MAIN SHOCKS

The instrumental catalogs involve not only the main shocks but also foreshocks and aftershocks. In estimating the earthquake hazard, generally, a Poisson model of earthquake occurrence is assumed. Therefore, the catalog in use must exhibit random space–time characteristics. Aftershocks and foreshocks show a major deviation from a Poisson process, and several methods have been suggested for the separation of aftershocks from the raw earthquake data (Savage, 1972; Gardner and Knopoff, 1974; Reasenberg, 1985; Davis and Frohlich, 1991; Molchan and Dmitrieva, 1992). Deleting aftershocks and other dependent events leads approximately to a Poisson or random data set for a better estimation of return periods of randomly occurring events, which is an important goal of seismic hazard studies. Declustering is the removal of foreshocks and aftershocks from the background seismicity

(Reasenberg, 1985). For seismicity rate studies (Wiemer and Wyss, 1994,1997), as well as hazard-related studies (Frankel, 1995), declustering is often considered necessary to achieve better results.

Knopoff (1964) introduced a declustering algorithm to count earthquakes in successive 10-day intervals and to prepare a histogram showing the traits of a Poisson distribution. Gardner and Knopoff (1974) introduced a procedure for identifying aftershocks within earthquake catalogs using distances in time and space. They also provided specific space–time distances as a function of the main shock magnitude to identify aftershocks. They ignored secondary and higher-order aftershocks (i.e., aftershocks of aftershocks). They also did not consider fault extension for earthquakes of larger magnitude by assuming circular spatial windows. The Reasenberg (1985) algorithm enables linking of aftershock triggering within an earthquake cluster. In this approach, the largest earthquake in a cluster is identified as the main shock. Another crucial development with this method is that the space–time distance is based on Omori's law (for its temporal dependence): as the time from the main shock increases, the time one must wait for the next aftershock increases in proportion (Stiphout et al., 2010).

The dependent shocks that fall within the space and time intervals of the main shock were eliminated to obtain a data set of main shocks that were assumed to show a Poisson distribution. The declustering was done following the algorithm developed by Gardner and Knopoff (1974) and modified by Uhrhammer (1986). A space and time window was used to remove foreshocks and aftershocks, as described in the following equations:

$$\text{Distance} = e^{-1.024 + 0.804 \times M_W} \qquad \text{Time} = e^{-2.87 + 1.235 \times M_W} \qquad (2.6)$$

Out of 203,448 events in the raw catalog, 75.3% were found to be dependent events; the remaining 50,317 events were identified as main shocks, of which 27,146 events were of $M_W \geq 4$. The number of earthquake events in the declustered catalog for different magnitude ranges is shown in Table 2.3 ($M_W \geq 4$). The distribution of earthquake events ($M_W \geq 4$) in the declustered catalog is shown in Figure 2.6. The details of all earthquake events compiled from 250 BC to 2010 AD homogenized in unified Moment magnitude scale is available at https://bit.ly/2w4EyDk, under Downloads/Updates tab. The details of major earthquakes in the region are listed in Table 2.4. Temporal changes of instrumental seismicity of both clustered and declustered catalogs are shown in Figure 2.7. In recent years, a rapid increase is seen in the number of earthquakes, which reflects the capability of advanced seismic recording instruments to record even smaller magnitude earthquakes in India.

Table 2.3 Statistics of earthquake events in the declustered catalog

Magnitude (M_W)	No. of events
4–4.9	16,079
5–5.9	9879
6–6.9	1036
7–7.9	129
8–9	22

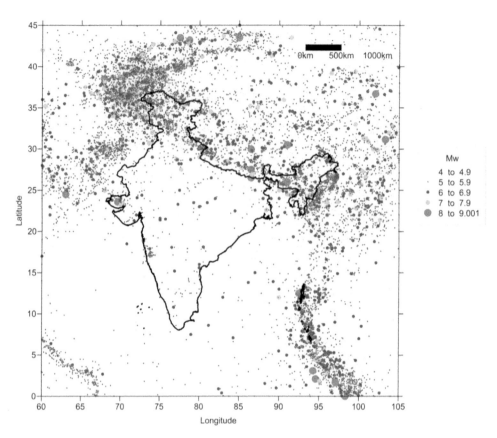

Figure 2.6 Distribution of earthquake events (main shocks only) in and around India (Kolathayar et al., 2012)

Table 2.4 Details of major earthquakes ($M_w > 7.5$) occurred in and around India (Kolathayar and Sitharam, 2012)

Year	Month	Date	Longitude (°)	Latitude (°)	M_w
1668	5	1	68.0	25.0	7.6
1737	5	11	88.4	22.6	7.7
1816	5	26	86.5	30.0	8
1819	6	16	69.6	23.6	8.3
1833	8	26	86.5	27.5	8.0
1897	6	12	91.0	26.0	8.1
1902	8	22	77.0	40.0	8.5
1902	8	30	71.0	37.0	7.7
1905	4	4	76.0	33.0	7.8
1908	10	23	70.5	36.5	7.6
1908	12	12	97.0	26.5	8.2
1911	7	4	70.5	36.5	7.6
1916	8	28	81.0	30.0	7.7
1918	7	8	91.0	24.5	7.6
1921	11	15	70.5	36.5	8.1

Year	Month	Date	Longitude (°)	Latitude (°)	M_w
1931	1	27	96.8	25.6	7.6
1932	12	25	96.5	39.2	7.6
1934	1	15	86.5	26.5	8.1
1937	1	7	98.0	35.5	7.6
1941	6	26	92.5	12.5	8.5
1947	3	17	99.5	33.0	7.7
1949	3	4	70.6	36.6	7.7
1950	8	15	96.5	28.6	8.6
1951	11	18	91.0	30.5	8.0
1956	6	9	69.1	34.3	7.6
1965	3	14	70.8	36.6	7.8
1983	12	30	72.0	34.5	7.7
1988	11	6	99.6	22.8	7.6
1997	11	8	87.325	35.069	7.6
2001	1	26	70.232	23.419	7.7
2001	11	14	90.541	35.946	7.8
2004	12	26	94.26	3.09	9
2005	10	8	73.588	34.539	7.6

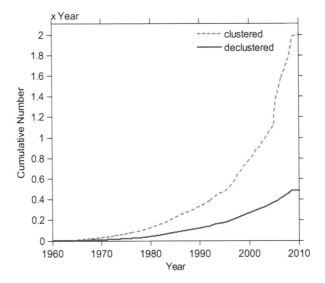

Figure 2.7 Cumulative number of earthquakes in the instrumental part of the catalog

2.5 COMPLETENESS ANALYSIS

The completeness of the catalog was analyzed using the method suggested by Stepp (1972), which provides the time interval in which the magnitude range is homogeneous. In the method proposed by Stepp (1972), the sample mean is inversely proportional to the number of observations in the sample. The earthquake sequence is modeled based on a Poisson distribution for obtaining an accurate estimate of the variance of the sample mean. If $x_1, x_2, x_3, \ldots \ldots x_n$ is the number of events per unit time interval, then the unbiased estimate of the mean per unit interval for this sample is given by:

$$\lambda = \frac{1}{n} \sum_{i=1}^{n} X_i \tag{2.7}$$

where n is the unit time interval, and its variance is $\sigma_\lambda^2 = \frac{\lambda}{n}$. When the time interval is taken as one year, the standard deviation of the above equation becomes $\sigma_\lambda = \frac{\sqrt{\lambda}}{\sqrt{T}}$; where T is the sample length. If this process is assumed to be stationary, it can be concluded that the standard deviation behaves as $\frac{1}{\sqrt{T}}$ in the subinterval of the complete years of the sample. Hence, during this period the mean rate of occurrence in a magnitude class will be constant.

As a first step for the evaluation of the completeness period, the number of earthquakes reported during each decade for the given magnitude ranges was evaluated. These results are given in Table 2.5. The plot showing the variation of σ_λ with time is given in Figure 2.8. The earthquake data are considered complete as long as the variation is along the $\frac{1}{\sqrt{T}}$ line. The completeness periods for different ranges of magnitude are given in Table 2.6. The completeness period of the smaller magnitude ranges will be low, as the region was not well instrumented in the past. Hence, lots of smaller magnitude events were not recorded correctly. However, the probability of events being recorded rises with increases in earthquake magnitude. That is the reason for the increasing trend in the completeness period with the increase in magnitude. The seismicity parameters have to be evaluated based on the complete part of the catalog.

2.6   DEVELOPMENT OF SEISMOTECTONIC MAP FOR INDIA

The compilation and integration of all available data on geological, geophysical, and seismological attributes for the entire country are required for the proper evaluation of seismicity in different tectonic regions. In this context, well-defined and well-documented seismic sources are published in the *Seismotectonic Atlas – 2000* (Dasgupta et al., 2000) published by the Geological Survey of India (GSI). The GSI has compiled all the available geological, geophysical, and seismological data for India. The atlas contains 43 maps covering India and adjoining areas, with all available data related to earthquakes. It is a multithematic database comprising 43 maps (presented in 42 sheets) covering India and adjacent regions of neighboring countries on a 1:1,000,000 scale. Various details regarding geophysical, structural, seismicity, and geothermal data relevant to seismotectonic activity are included in the *Seismotectonic Atlas* (SEISAT).

Table 2.5 Number of earthquakes per decade

Time (Years)	Magnitude Range											Total
	4.0–4.49	4.5–4.99	5.0–5.49	5.5–6	6.0–6.49	6.5–6.99	7.0–7.49	7.5–7.99	8.0–8.49	8.5–8.99	9.0–9.5	
2001–2010	1202	3191	1413	386	80	37	5	6	1	1	1	20414
1991–2000	1553	4468	1886	664	181	50	17	2	0	0	0	15533
1981–1990	398	3345	2028	646	127	68	15	4	0	0	0	8381
1971–1980	125	1405	1150	369	88	32	10	4	0	1	0	3683
1961–1970	17	279	496	258	57	18	2	2	0	1	0	1180
1951–1960	3	5	66	70	44	12	3	2	0	1	0	208
1941–1950	0	1	56	46	30	8	1	4	2	1	0	149
1931–1940	0	14	73	38	49	8	3	5	0	2	0	193
1921–1930	0	3	82	41	44	5	6	0	0	2	0	183
1911–1920	0	1	9	9	8	8	0	5	1	0	0	41
1901–1910	0	2	6	1	4	8	2	4	1	3	0	31
1891–1900	1	0	0	3	2	2	0	1	0	1	0	14
1881–1890	1	0	2	3	2	3	4	0	0	0	0	22
1871–1880	0	1	5	2	0	1	1	0	1	0	0	16
1861–1870	9	3	6	10	3	3	1	1	0	1	0	50
1851–1860	8	9	4	3	1	1	0	0	0	0	0	39
1841–1850	3	2	7	9	6	4	1	0	0	0	0	43
1831–1840	2	2	2	6	1	1	1	0	0	1	0	16
1821–1830	6	7	9	5	4	0	0	0	0	0	0	32
1811–1820	2	1	4	0	3	0	0	1	0	2	0	14
1801–1810	3	1	2	1	1	1	1	0	0	0	0	11
1791–1800	0	0	1	0	0	1	0	0	0	0	0	4
1781–1790	0	0	0	0	0	2	0	1	0	0	0	3
1771–1780	0	0	0	0	0	0	1	0	0	0	0	1
1761–1770	1	0	0	0	2	2	0	0	0	0	0	5

(Continued)

Table 2.5 (Continued)

Time (Years)	Magnitude Range											Total
	4.0–4.49	4.5–4.99	5.0–5.49	5.5–6	6.0–6.49	6.5–6.99	7.0–7.49	7.5–7.99	8.0–8.49	8.5–8.99	9.0–9.5	
1751–1760	1	3	0	0	1	1	0	0	0	0	0	10
1741–1750	0	0	0	1	0	0	0	0	0	0	0	1
1731–1740	0	0	0	0	0	1	0	1	0	0	0	2
1721–1730	0	0	0	0	0	0	0	0	0	0	0	0
1711–1720	0	0	0	0	0	0	0	2	0	0	0	2
1701–1710	0	0	0	0	0	0	0	0	0	0	0	1
1691–1700	0	0	0	0	0	0	0	0	0	0	0	0
1681–1690	0	0	0	0	0	1	0	0	0	0	0	2
1671–1680	0	0	1	0	0	1	0	0	0	0	0	2
1661–1670	0	0	0	0	3	1	0	1	0	0	0	5
1651–1660	0	0	0	0	0	0	0	0	0	0	0	0
1641–1650	0	0	0	0	0	0	0	0	0	0	0	0
1631–1640	0	0	0	0	0	0	0	0	0	0	0	0
1621–1630	0	0	0	0	0	0	0	0	0	0	0	0
1611–1620	0	0	0	0	1	0	1	0	0	0	0	2
1601–1610	0	0	0	0	1	1	0	0	0	0	0	2
1591–1600	0	0	0	0	0	0	0	0	0	0	0	1
1581–1590	0	0	0	0	0	0	0	0	0	0	0	0
1571–1580	0	0	0	0	1	0	0	0	0	0	0	1
1561–1570	0	0	0	0	0	1	0	0	0	0	0	0
1551–1560	0	0	0	0	0	0	0	0	0	0	0	2
1541–1550	0	0	0	0	0	0	0	0	0	0	0	0
1531–1540	0	0	0	0	0	0	0	0	0	0	0	0
1521–1530	0	0	0	0	0	0	0	1	0	0	0	0
1511–1520	0	0	0	0	1	0	0	0	0	0	0	2
1501–1510	0	0	0	0	0	0	1	0	0	0	0	1

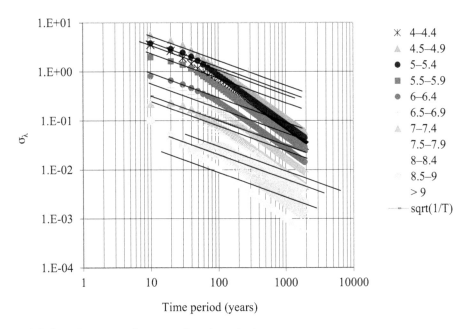

Figure 2.8 Completeness analysis using Stepp's method

Table 2.6 Catalog completeness for different ranges of magnitude

Magnitude range	Catalog is complete for
$4.0 < M_w < 4.5$	40 years
$4.5 < M_w < 5.0$	45 years
$5.0 < M_w < 5.5$	46 years
$5.5 < M_w < 6.0$	50 years
$6.0 < M_w < 6.5$	80 years
$6.5 < M_w < 7.0$	100 years
$7.0 < M_w < 7.5$	115 years
$7.5 < M_w < 8.0$	120 years
$8.0 < M_w < 8.5$	140 years
$8.5 < M_w < 9.0$	140 years
$9.0 < M_w < 9.5$	140 years

2.6.1   Scanning of maps

A map is the representation of a geographic area on a piece of paper or canvas. In order to study maps or aerial images it is necessary to convert them into digital form. The 42 sheets in SEISAT published by the GSI were scanned separately at a resolution of 300 dots per inch (dpi) to obtain high-quality digital images.

2.6.2   Georeferencing and digitization

The power of geographic information systems (GIS) technology makes creating and modifying maps a straightforward task. Many different attributes can be highlighted using colors and symbols in GIS. Application of GIS techniques allowed for the inserting, extracting, handling, managing, and analyzing of data for the zoning of seismicity. The capability of GIS to store and process data and images makes it very valuable in the field of seismic studies. MapInfo Professional Version 6.0 was used for georegistration and digitization of the scanned maps. A new table of infinite dimension was created in MapInfo, and each sheet was brought into the table. The coordinates were fixed based on the latitudes and longitudes of the region in the map.

All faults, shear zones, lineaments, and other seismic sources were digitized using different legends in each map in different layers. Lastly, all 42 digitized digital images were merged to form the whole map of India with all tectonical features in it. Most of the sources hidden in the ocean are not quantified, and hence are not identified in the present map. Figure 2.9 shows the map of India with all identified structural symbols. The complete homogenized earthquake events data were superimposed on this source map to get the final digital seismotectonic map of India with all the seismic sources and earthquake events (Fig. 2.10).

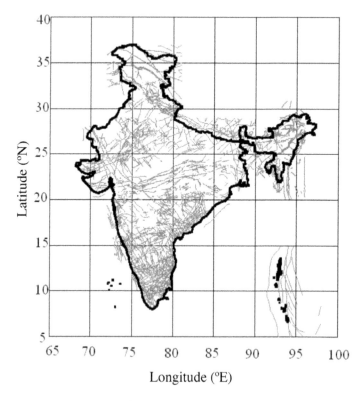

Figure 2.9 Linear seismic sources identified in India

Figure 2.10 Digitized seismotectonic map of India showing linear sources and events of $M_w \geq 4$

2.7 SEISMIC SOURCE MODELS

Another essential step in seismic hazard analysis is the identification of vulnerable seismic sources. The types of seismic sources considered in the present study are linear seismic sources, point sources, a gridded seismicity source model, and areal sources.

2.7.1 Linear seismic sources

The movement of the Indian plate in the northeast direction and its collision with the Eurasian plate have resulted in the shortening of the Indian plate (Mahajan et al., 2010). The collision has created three major south-verging thrust faults that comprise the most critical tectonic features in the Himalayas that accommodate the convergence across the Himalayan belt (Rajendran and Rajendran, 2011). The three major thrust faults along the Himalayas are (1) the Main Central Thrust (MCT), which is the northernmost of the three; (2) the Main Boundary Thrust (MBT), which is along the southern edge of the Lesser Himalaya; and (3) the Himalayan Frontal Thrust (HFT), which is the southernmost thrust fault. The northernmost and the oldest of this thrust system is the MCT, which marks the contact between the Higher and the Lesser Himalayas (Gansser, 1964). The north-dipping thrust faults of the MBT, which lies south of the MCT, separate the pre-Tertiary Lesser Himalayan sediments from the Tertiary and Quaternary sub-Himalayan sediments (Rajendran and Rajendran, 2011). The southernmost and the youngest of the three thrusts is the HFT, which occurs as a discontinuous range of slopes of cutting Quaternary fluvial terraces and alluvial fans (Kumar et al., 2001; Nakata, 1989; Valdiya, 1992).

The major tectonic features associated with past seismic activity in India are well known. One of the best documents listing the linear seismic sources in India and adjoining areas is SEISAT (Dasgupta et al., 2000) published by the GSI. It was prepared through extensive studies using remote sensing techniques and geological explorations. SEISAT maps are available in A_0-size sheets at a 1:1,000,000 scale, and each map covers an area of $3° \times 4°$. SEISAT contains the details of the faults, lineaments, and shear zones, in addition to the geological features in India and adjoining areas. It is considered an authentic reference manual for identifying seismic sources by various researchers, including Iyengar and Ghosh (2004) for Delhi; Nath et al. (2006) for microzonation of the Sikkim Himalayas; Raghu Kanth and Iyengar (2006) for Mumbai; Boominathan et al. (2008) for Chennai; Anbazhagan et al. (2009) for Bangalore; and Vipin et al. (2009) for south India. Thus, in the present study, SEISAT was taken as the main reference for identifying linear seismic sources. The linear seismic sources mentioned by Ganesha Raj and Nijagunappa (2004) and Gupta (2006) were also considered for the analysis. The digitization of linear sources was explained in section 2.6. All 43 sheets of SEISAT were scanned using a high-resolution scanner to digitize these maps. The individual maps were then georeferenced using MapInfo Professional version 6.0. After georeferencing, the individual maps (images) were combined to form a complete map of India and adjoining areas. From this map, the tectonic features were carefully extracted, and these extracted data were kept as a separate layer.

The declustered earthquake data with M_W 4.0 and above were superimposed on the digitized map of extracting tectonic features. To characterize the seismic sources, the maximum reported magnitude and number of earthquake events associated with each of these sources were noted. If any event fell within a 15-km radius of two or more sources, it was assumed that the event was associated with the source nearest to the event. Some earthquake events did not fall along any of the identified faults. In those cases, the linear source model alone may not be able to give the correct picture of seismic hazard levels. To overcome this limitation, three alternative source models were adopted – point sources, smoothed gridded sources, and areal sources. These are widely accepted source models for evaluation of seismic hazard for regions where the sources are not delineated.

2.7.2   Point sources

Point sources are the sources that are spread over an area to be used in deterministic seismic hazard analysis. Even though linear seismic sources were identified from SEISAT and other literature, it can be seen that many earthquake events are not associated with any of the identified linear seismic sources. When the seismic hazard analysis is done using the linear seismic sources alone, the effects of such earthquake events will not be considered. This, in turn, will lead to an erroneous interpretation of the seismic hazard. In those cases, the modeling of seismic sources as point sources, the sources which are spread over an area, is a better option. For identification of these sources, the study area was divided into grids of size $0.2° \times 0.2°$ and the maximum reported magnitude within each cell was assigned to the center of that cell.

After assigning the maximum magnitude to the center of each grid for the entire study area, these observed maximum magnitudes were smoothed using a centered smoothing window. The smoothing is done to account for the source dimension and location errors (Costa et al., 1993; Panza et al., 1999). Depending on the magnitude of the earthquake, the rupture length will vary, and the smoothing will take this aspect into account. The smoothing windows selected had radii of 0.2°, 0.4°, and 0.6° (radii of grids 1, 2, and 3) for magnitude ranges of 4.0–4.9, 5.0–5.9, and ≥ 6 respectively. Costa et al. (1993) adopted a similar approach for smoothing earthquake magnitudes. However, they used a smoothing window with a constant radius of 0.6° for all the magnitude ranges.

The smoothing process of the study area was done from the extreme southwestern grid point (bottom-left grid). The center of the smoothing window was kept at that grid point, and its radius was selected based on the magnitude of earthquakes in that grid. While smoothing, the earthquake magnitudes were assigned only to those cells that had a number of earthquakes equal to or greater than a threshold value. Once the smoothing of that particular grid point was over, the center of the smoothing window shifted to the adjoining grid point on the same row, and the same process was repeated. A new program for this purpose was developed in MATLAB, and the smoothing of the earthquake magnitudes was done using this program. During the smoothing process, the program kept in memory all the original magnitudes in each of the grid points. This was essential because while smoothing it is not permissible to overwrite a larger magnitude with a smaller one. After completing the smoothing for a particular row, the center of the smoothing window shifted to the next row, and the process was repeated until the smoothing of the entire study area was complete.

2.7.3 Gridded seismicity source model

The gridded seismicity source model (Frankel, 1995; Woo, 1996; Martin et al., 2002) is based on the seismic activity rate obtained from the earthquake catalog. It is one of the most widely adopted methods to model seismic sources for regions in the absence of clearly identified seismic sources. Some seismic hazard studies that have considered the zoneless approach for source identification are Wahlstrom and Grunthal (2000), Lapajne et al. (2003), Vilanova and Fonseca (2007), Jaiswal and Sinha (2007), Kalkan et al. (2009), and Menon et al. (2010). In this method, the study area is divided into grids, and the number of earthquakes that have a magnitude higher than a cutoff magnitude in each grid are counted. This gives the activity rate for that particular grid cell. Based on this value, the recurrence rates for different magnitude intervals can be calculated, and these values can be smoothed using a Gaussian function to get the final activity rate for each grid cell. The uncertainty involved in estimating the location of the earthquake event and the size of the seismic source can be accounted for by this smoothing.

When considering these sources, the first step is the selection of the grid size and the cutoff magnitude (M_{cut}). After selecting a suitable grid size, the number of earthquake events of magnitude greater than or equal to M_{cut} is calculated for each grid cell, and this represents the maximum likelihood estimate of the total number of earthquakes (equal to or greater than M_{cut}) for that grid cell. In the present study, the value of M_{cut} was taken as 4.0 to eliminate the effect of rock bursts (blasting). Moreover, seismic events with magnitude less than 4.0 may not cause much damage. Based on this value, the recurrence rates for different magnitude intervals were calculated. These values were smoothed using a Gaussian function to get the final corrected values for each grid. This smoothing was performed to account for the uncertainty associated with the location of earthquake events:

$$\hat{n}_i = \frac{\sum_j n_j e^{-\Delta_{ij}^2/c^2}}{\sum_j e^{-\Delta_{ij}^2/c^2}} \tag{2.8}$$

where n_j is the number of earthquakes in the jth grid cell; $\hat{n}_i$ is the smoothed number of earthquakes in the ith cell; c is the correlation distance, to account for the location uncertainties; and Δ_{ij} is the distance between the ith and j^{th} cells.

2.7.4 Areal sources

For hazard estimation using areal sources, the territory under study should first be divided into seismic sources such that within a seismic source the earthquake-occurrence process is independent. For each seismic source, separate earthquake sub-catalogs have to be generated and magnitude exceedance rates need to be estimated using statistical analysis of these sub-catalogs. These rates are the number of earthquakes per unit time, in which a specific magnitude is exceeded, and they represent the seismicity of the source (Ordaz et al., 2007).

These zones allow for local variations in seismicity characteristics for the other two types of source models: linear sources and the zoneless approach (e.g., changes in a, b values, M_{max}, etc.). With this method, a spatial integration process can be carried out to account for all possible focal locations with an assumption that, within a seismic source, all points are equally likely to be an earthquake focus. CRISIS software (Ordaz et al., 2007) is a potential tool used to model areal sources and to estimate the seismic hazard with polygon-dipping areas. CRISIS assumes that, within a source, seismicity is evenly distributed by unit area; to correctly account for this modeling assumption, CRISIS performs spatial integration by subdividing the original sources. Once subdivided into subsources, CRISIS assigns to a single point all the seismicity associated with a subsource, and then the spatial integration adopts a summation form.

2.8 SUMMARY

An updated earthquake catalog that is uniform in moment magnitude and fairly complete at the $M_W \geq 4.8$ level has been prepared for India and adjoining areas for the period until 2010. Region-specific magnitude scaling relations were established for the study region, which facilitated the generation of a homogenous earthquake catalog. By carefully converting these original magnitudes to unified M_W magnitudes, a major obstacle was removed for consistent assessment of seismic hazards in India. The earthquake catalog was declustered to remove the aftershocks and foreshocks. Out of 203,448 events in the raw catalog, 75.3% were found to be dependent events; the remaining 50,317 events were identified as main shocks, of which 27,146 events were of $M_W \geq 4$. A completeness analysis of the catalog was carried out to estimate the completeness periods of different magnitude ranges. The details of the earthquake events are available at https://goo.gl/VXFiJy.

Chapter 3

Deterministic seismic hazard assessment

3.1 INTRODUCTION

The most important seismic hazard parameters required to delineate seismic zones are the peak ground acceleration (PGA) and the spectral acceleration (Sa). The design ground motion at a site is determined by conducting a seismic hazard analysis (SHA). The two approaches for evaluation of seismic hazard are deterministic seismic hazard analysis (DSHA) and probabilistic seismic hazard analysis (PSHA). DSHA is a simple methodology that uses geology and seismic history to identify earthquake sources and to interpret the strongest earthquake each source is capable of producing, regardless of time (Krinitzsky, 2005). This is termed the maximum credible earthquake (MCE), the largest earthquake that can reasonably be expected. The MCE is the largest earthquake that appears possible along a recognized fault under presently known or presumed tectonic activity (USCOLD, 1995). DSHA calculates the hazard parameters based on the geological facts, and it is simple and transparent; PSHA is based on numerical calculations and probability theory (Krinitzsky, 2003).

3.2 METHODOLOGY

DSHA considers a particular earthquake scenario, either realistic or assumed. It uses known seismic sources that are near the site and available historical seismic and geological data to generate distinct models of ground motion at the site. The earthquakes are assumed to occur on the source closest to the site. DSHA requires three input details to evaluate the seismic hazard: the earthquake source, the controlling earthquake at the source, and an attenuation relation. A schematic diagram of the different steps involved in DSHA is provided in Figure 3.1. In DSHA, the controlling earthquake is assumed to act along the source the shortest distance from the site. The uncertainties involved in the earthquake magnitude or location are not taken into account, and this method will give an upper-bound value for the ground motion. Hence, DSHA is adopted in the evaluation of seismic hazard for critical structures such as nuclear power plants, big dams, bridges, hazardous waste contaminant facilities, etc. The results obtained from deterministic analysis can be used as a cap for the probabilistic analysis.

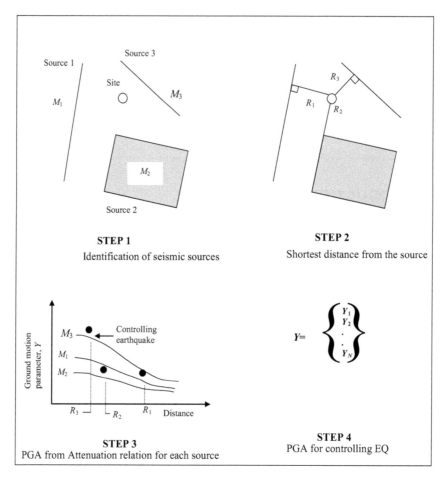

Figure 3.1 The various steps involved in DSHA

Source: Kramer (1996).

3.3   ESTIMATION OF HAZARD FOR INDIAN SUBCONTINENT

Given the significant advances made in understanding the seismicity and seismotectonics of Indian subcontinent, this section presents seismic hazard evaluation of the Indian landmass (covering 6°to 38°N and 68°to 98°E) based on a state-of-the-art DSHA using different source models and attenuation relations. The DSHA is performed with currently available data and their best possible scientific interpretation using an appropriate instrument such as the logic tree to explicitly account for epistemic uncertainty by considering alternative models (source models and GMPEs). As noted, with DSHA, the controlling earthquake is assumed to act along the source at the shortest distance from the site. For calculating the seismic hazard values, the entire study area was divided into grids of size 0.1° × 0.1°, and the hazard values at the center of each grid cell were calculated by considering all the seismic sources within a radius of 500 km. At a particular grid point, the expected PGA value was estimated for each of these sources by considering the closest possible distance and maximum earthquake associated with that source. When calculating the minimum distance, the focal depth and the curvature of the Earth were also considered. The distance type specific to each attenuation relation was considered in the analysis in estimating PGA using different attenuation models. The source that gave the highest value of the PGA has the maximum earthquake potential, and that PGA was considered for that grid point. This exercise was repeated for each grid point using all the attenuation relations. Two different types of source models, namely, linear sources and point sources, were considered in the analysis, the results of which were combined in a logic tree framework. The details of the source models used are discussed in Chapter 2.

3.3.1 Attenuation models

The seismicity of the Indian subcontinent is spatially varied and complex as it embraces various tectonic zones with different attenuation characteristics. North and northwest India are active tectonic regions with shallow crustal seismic activity. The seismic activity in the Indo-Myanmar subduction zone in northeast India is because of intraslab subduction earthquakes. The Andaman and Nicobar Island region also comes under a subduction zone, but with earthquakes of an interface nature. South and central India are stable continental regions with low to moderate seismicity. Different attenuation relations should be used in these regions. Hence, for the selection of ground motion prediction equations (GMPEs), the study area was divided into four categories: active tectonic shallow crustal region, stable continental region, subduction intraslab region, and subduction interface region.

In the present analysis, three different GMPEs were used to model the attenuation properties of the plate boundary region, shield region, interface subduction zones, and intraslab subduction zones. The relations used for the shield region are from Raghu Kanth and Iyengar (2007), Atkinson and Boore (2006), and Campbell and Bozorgnia (2003). Of these, the relation by Raghu Kanth and Iyengar (2007) was developed for the peninsular Indian shield regions. Attenuation relations given by Campbell and Bozorgnia (2003) and Atkinson and Boore (2006) were developed for eastern North America (ENA). Based on a study of aftershocks of the Bhuj earthquake, Cramer and Kumar (2003) concluded that the ground motion attenuation in ENA and the peninsular Indian shield are comparable. The similarity of the regional tectonics of ENA and peninsular India also has been noted by Bodin et al. (2004).

The GMPEs used for active tectonic regions are Sharma et al. (2009),Boore and Atkinson (2008), and Akkar and Bommer (2010). Of these, the relation suggested by Sharma et al. (2009) was developed for Himalayan regions of India. Sharma et al. (2009) used data from the Himalayan and Zagros regions on the premise that the seismotectonics of the two regions have a considerable similarity (Ni and Barazangi, 1986). In contrast, the relation by Boore and Atkinson (2008) was developed for active tectonic regions around the world, and that offered by Akkar and Bommer (2010) was developed for the active tectonic regions of Europe and the Middle East.

For the Indo-Myanmar subduction zone, we used the attenuation relations suggested by Gupta (2010), Zhao et al. (2006), and Lin and Lee (2008), as all three are capable of predicting ground motion from intraslab subduction earthquakes. For the subduction zone with interface earthquakes, the GMPEs used were of Lin and Lee (2008), Atkinson and Boore (2003), and Gregor et al. (2002). The GMPE by Gupta (2010) was explicitly developed for

Table 3.1 Attenuation relations used for different tectonic provinces of the Indian region

Shield region	Active tectonic regions	Subduction zone (Intraslab region)	Subduction zone (Interface region)
1 Raghu Kanth and Iyengar (2007)	Sharma et al. (2009)	Gupta (2010)	Lin and Lee (2008)
2 Atkinson and Boore (2006)	Boore and Atkinson (2008)	Lin and Lee (2008)	Atkinson and Boore (2003)
3 Campbell and Bozorgnia (2003)	Akkar and Bommer (2010)	Zhao et al. (2006)	Gregor et al. (2002)

the Indo-Myanmar subduction zone, whereas those given by Zhao et al. (2006) and Lin and Lee (2008) were designed for the subduction regions (both intraslab and interface) of Japan and Taiwan, respectively. Attenuation relations given by Gregor et al. (2002) and Atkinson and Boore (2003) were developed for the Cascadia subduction zone. The GMPE models for different seismotectonic provinces considered in the present study are summarized in Table 3.1.

Attenuation relation by Akkar and Bommer (2010)

$$\log(PSA) = b_1 + b_2 M + b_3 M^2 + (b_4 + b_5 M)\log\sqrt{R^2 + b_6^2} + b_7 S_s + b_8 S_A + b_9 F_N$$
$$+ b_{10} F_R + \varepsilon\sigma$$

where S_s and S_A take the value of 1 for soft ($Vs^{30} < 360$ ms^{-1}) and stiff soil sites, otherwise zero; rock sites are defined as having $Vs^{30} > 750$ ms^{-1}; similarly F_N and F_R take the value of unity for normal and reverse faulting earthquakes, respectively, otherwise zero; e is the number of standard deviations.

Attenuation relation by Atkinson and Boore (2003)

$$\log(Y) = c_1 + c_2 M + c_3 h + c_4 R - g\log R + C_5 slS_c + c_6 slS_D + c_7 slS_E$$

Here, Y represents the random horizontal component of the peak ground acceleration, M the moment magnitude (limited to 8.5 for interface and to 8.0 for in-slab events with larger magnitude), h the focal depth in kilometers (limited to 100 km for deeper events), R a distance metric with near-source saturation effects taken into account, and g the geometric attenuation factor.

Attenuation relation by Atkinson and Boore (2006)

$$\log(PSA) = c_1 + c_2 M + c_3 M^2 + (c_4 + c_5 M)f_1 + (c_6 + c_7 M)f_2 + (c_8 + c_9 M)f_0 + c_{10} R_{cd} + S$$

where, $f_0 = \max\left[\log\left(\frac{R_0}{R_{cd}}\right), 0\right] f_1 = \min(\log R_{cd}, \log R_1)$, $f_2 = \max\left[\log\left(\frac{R_{cd}}{R_2}\right), 0\right] R_0 = 10$, $R_1 = 70$, $R_2 = 140$, and $S = 0$ for hard rock sites.

Attenuation relation by Boore and Atkinson (2008)

$$\ln(Y) = F_M(M) + F_D(R_{JB}M) + F_S(V_S^{30}, R_{JB}, M) + \varepsilon\sigma_T$$

In this equation, F_M, F_D, and F_S represent the magnitude scaling, distance function, and site amplification, respectively. M is moment magnitude, R_{JB} is the Joyner-Boore distance (defined as the closest distance to the surface projection of the fault, which is approximately equal to the epicentral distance for events of $M < 6$).

Attenuation relation by Gupta (2010)

$$\log(Y) = c_1 + c_2M + c_3h + c_4R - g\log R + C_5 slS_c + c_6 slS_D + c_7 slS_E$$

Here, Y represents the random horizontal component of the peak ground acceleration, M is the moment magnitude (limited to 8.5 for interface and 8.0 for in-slab events with larger magnitude), h the focal depth in kilometers (limited to 100 km for deeper events), R a distance metric with near-source saturation effects taken into account, and g the geometric attenuation factor.

Attenuation relation by Sharma et al. (2009)

$$\log A = b_1 + b_2M + b_3 \log \sqrt{R_{JB}^2 + b_4^2} + b_5S + b_6H$$

where b_1, b_2, b_3, b_4, b_5, and b_6 are the regression coefficients; A is the spectral acceleration in terms of ms^{-2}; S is 1 for a rock site and 0 otherwise; and H is 1 for a strike-slip mechanism and 0 for a reverse mechanism.

Attenuation relation by Lin and Lee (2008)

$$\ln(PGA) = -2.5 + 1.205M - 1.905\ln(R + 0.516e^{0.6325M}) + 0.0075H + 0.275Z_t$$

Attenuation relation by Zhao et al. (2006)

$$\log(y_{i,j}) = AMw_i + bx_{i,j} - \log_e(r_i, j) + e(h - h_c)\delta_h + F_R + S_I + S_S + S_{SL} \log_e(x_{i,j})$$
$$+ C_k + \zeta_{i,j} + \eta_i$$

where y is the PGA, M_W the moment magnitude, x the source to site distance, and h the focal depth in kilometers.

Attenuation relation by Campbell and Bozorgnia (2003)

$$\ln Y = c_1 + f_1(M_W) + c_4 \ln \sqrt{f_2(M_W, r_{seis}, S} + f_3(F) + f_4(S) + f_5(HW, F, M_W, r_{seis}) + \varepsilon_r$$

where f_1, f_2, f_3, f_4, and f_5 denote magnitude scaling characteristics, distance scaling characteristics, the faulting mechanism, the far source effect, and the effect of the hanging wall, respectively.

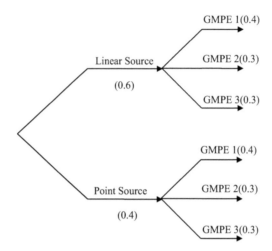

Figure 3.2 Parameters and weighting factors adopted in the logic tree

3.3.2   Logic tree structure

Many uncertainties are involved in the models used for seismic hazard assessment, and this makes the selection of a seismic hazard model difficult. The use of the logic tree approach allows characterization of epistemic uncertainties in various models by including alternative models in the analysis (Budnitz et al., 1997; Stepp et al., 2001; Bommer et al., 2005). The logic tree consists of a series of nodes and branches, and these branches denote different models (hypotheses). A subjective weighting, based on engineering judgment, can be given to each of these branches depending on the likelihood of it being correct. The weight for all the branches at a particular node should be equal to unity. The weight of the terminal branch of the logic tree can be obtained by multiplying the weights of all the branches leading to it. The present study considers two types of source models and three different attenuation relations for various tectonic provinces in the study area. These different models were combined using the logic tree with different weights (Fig. 3.2).

3.4   DISCUSSIONS

The outcome consists of seismic hazard contour maps of India for the horizontal component of ground motion for different structural periods (PGA, spectral acceleration at 0.1 s and 1.0 s) on bedrock conditions. Hazard maps were produced for horizontal ground motion at bedrock level (shear wave velocity ≥ 3.6 kms⁻¹) and compared with the seismic hazard zoning maps by the Indian seismic standards (BIS-1893 part 1, 2002) and Parvez et al. (2003). The Indian code is based on intensity and geological data; it is not based on a scientific seismic hazard assessment. The PGA values obtained for the five most populous cities (megacities and metros) of India are given in Table 3.2. The spatial variation of PGA values obtained is shown in Figure 3.3. The spectral acceleration values obtained for the periods 0.1 s and 1 s are shown in Figures 3.4 and 3.5. It can be seen that the seismic hazard is high along the plate boundary

Table 3.2 PGA values at rock level for the five most populous cities of India

Major cities	Location		PGA value (g)		Z/2 reported by BIS-1893 (2002)
	Longitude (°E)	Latitude (°N)	Present value	Parvez et al. (2003)	
Mumbai	72.82	18.90	0.27	Not reported	0.08
New Delhi	77.20	28.58	0.38	0.15 to 0.30	0.12
Bengaluru	77.59	12.98	0.13	Not reported	0.05
Kolkata	88.33	22.53	0.30	0.01 to 0.02	0.08
Chennai	80.25	13.07	0.1	Not reported	0.08

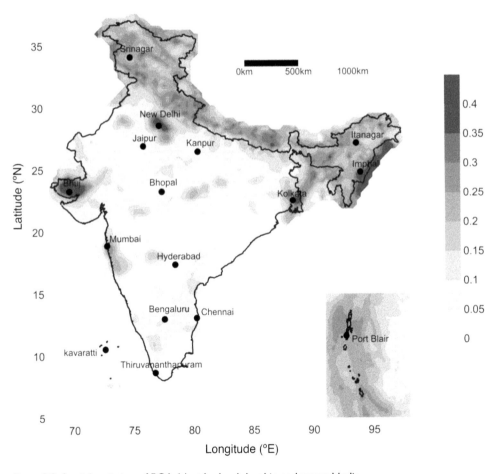

Figure 3.3 Spatial variation of PGA (g) at bedrock level in and around India

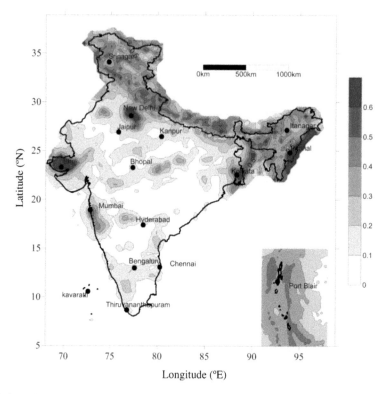

Figure 3.4 Spatial variation of PGA (g) for 0.1 s at bedrock level in and around India

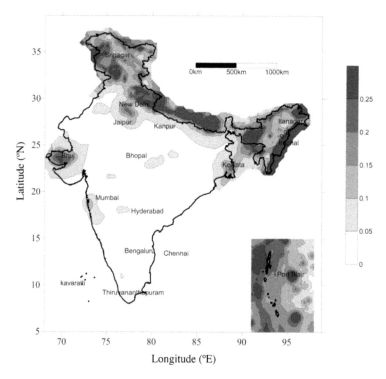

Figure 3.5 Spatial variation of PGA (g) for 1 s at bedrock level in and around India

regions like north and northeast India and the Andaman and Nicobar Islands. Along the shield region, the highest hazard is observed at the Bhuj region and in the Koyna region.

The PGA values for the plate boundary regions range from 0.3 g to 0.5 g, whereas for the shield region the values are less than 0.25 g except for the Kutch region in Gujarat. BIS 1893–1 (2002) delineates the country into four seismic zones and expresses the hazard in terms of zone factors. The design horizontal seismic coefficient Ah for a structure shall be determined by the following expression: $A_h = Z/2 \times I/R \times S_a/g$. Assuming I/R and S_a/g as equal to 1, $A_h = Z/2$. Hence, the obtained results are compared with the value of $Z/2$ (Table 3.2). This is a first-level comparison.

The PGA values obtained in the present study for Delhi, the Kutch region, and northeast India are in good agreement with those obtained by Parvez et al. (2003), whereas the present PGA values for many other parts of the country, especially the peninsular shield, are higher than those reported by them. The comparison of present PGA values with those of Parvez et al. (2003) for important cities is also shown in Table 3.2. Note that Parvez et al. (2003) have not reported the PGA values for many parts of the peninsular shield. Present PGA values for low seismic regions match with the values reported by Vipin and Sitharam (2011).

For comparison and evaluation, the PGA values were estimated without the use of the logic tree approach. The PGA values estimated using the linear source model and GMPEs given by Atkinson and Boore (2006), Sharma et al. (2009), Gupta (2010), and Gregor et al. (2002) are shown in Figure 3.6. The hazard values estimated using the point source model

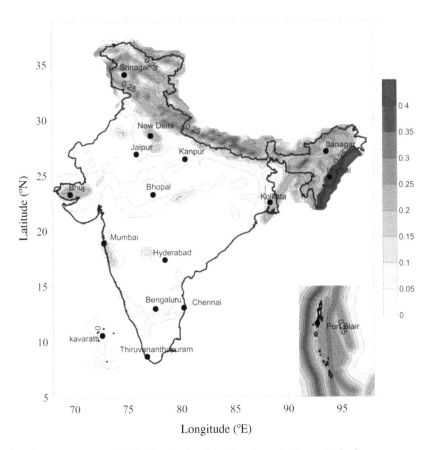

Figure 3.6 Spatial variation of PGA (g) at bedrock level estimated using only the linear source model

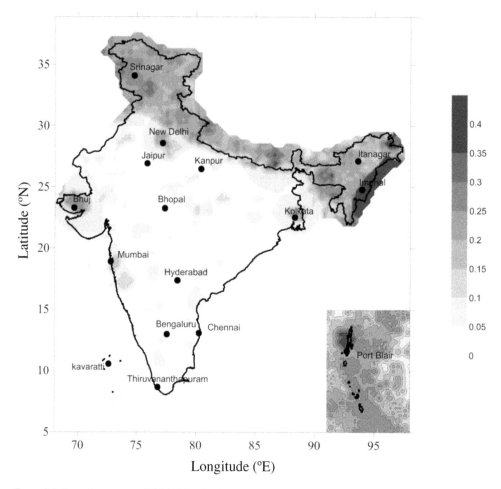

Figure 3.7 Spatial variation of PGA (g) at bedrock level estimated using only the point source model

and the same set of attenuation relations are shown in Figure 3.7. It is seen that the point source model predicts high hazard in and around the location of earthquakes, whereas the linear source model predicts the same along the whole length of a tectonic feature associated with an earthquake.

The uncertainties involved in earthquake magnitude or location are not taken into account in DSHA, and this method will give an upper-bound value for the ground motion. Hence, DSHA is adopted in the evaluation of seismic hazard for critical structures such as nuclear power plants, big dams, bridges, and hazardous waste contaminant facilities. The ground motion predicted from DSHA not only aids in the safe design of structures but will also help in determining the best locations for important structures such as nuclear power plants.

Chapter 4

Seismicity analysis and characterization of source zones

4.1 INTRODUCTION

The seismicity of a region can be characterized based on the seismicity parameters. Seismicity parameters are estimated based on the temporal distribution of different magnitude earthquake events in a region. This chapter describes the details of seismicity analysis for estimation of seismicity parameters for a region. Further, an attempt has been made to identify and characterize regional seismic source zones across the whole of India and adjoining area (5°–40°N and 65°–100°E). The tectonic framework of the Indian subcontinent, covering an area of about 3.2 million km², is complex and varied spatially. This necessitates identifying different regions of similar seismicity. Identification and characterization of seismic sources are essential inputs into seismic hazard analysis. Regional seismic source zones of homogeneous seismicity can be identified based on seismic event distribution, geology, and fault alignment. Separate catalogs need to be created for each zone, and seismicity parameters can be estimated for all the source zones with the maximum likelihood method based on the threshold magnitude.

4.2 SEISMICITY ANALYSIS

Several efforts have been made to determine seismicity parameters for the Indian region based on both the historical as well as the instrumental earthquake catalog (Shanker and Sharma, 1998; Iyengar and Ghosh, 2004; Raghu Kanth and Iyengar, 2006; Jaiswal and Sinha, 2007; Raghu Kanth, 2010). These researchers focused on some part of India or used a different methodology for homogenization of the catalog, as well as estimation of seismicity parameters.

The size distribution of earthquakes in a seismogenic source can often be adequately described over a broad range of magnitudes by a power law relationship. This was explained by Gutenberg and Richter (1944) using earthquake data from California. The commonly used form of the power law is given as log $N = a - bM$, where N is the cumulative number of earthquakes and a and b are constants. The parameter a describes the productivity of a volume, and b, the slope of the frequency-magnitude distribution (FMD), describes the relative size distribution of events. Spatial mapping of b values has been proven as a rich source of information about the seismotectonics of a region. The high-quality earthquake catalogs collected mainly over the past 20 years, and the availability of increased computing power, have enabled researchers to investigate spatial variations in b with high precision. The strong differences in *b* are merely a reflection of the heterogeneity of the earth that emerges on all scales, once suitable datasets become available (Wiemer and Wyss, 2002). The maximum likelihood method was used in the estimation of seismicity parameters. In this study, the effect of dependent events (aftershocks) on earthquake hazard parameters is also examined by using two kinds of catalogs: clustered and declustered.

4.2.1   Magnitude of completeness

The magnitude of completeness (M_c) is the lowest magnitude above which the earthquake recording is assumed to be complete. The magnitude of completeness is defined as the lowest magnitude at which 100% of the events in a space–time volume are detected (Rydelek and Sacks, 1989). Below this magnitude, a fraction of events is missed by the network because the events are either too small to be recorded by enough stations, or because they are below the magnitude of interest, or because they are mixed with the coda of a larger event, and therefore they passed undetected. To study the spatial variation of the seismicity parameters, the study area was divided into small grids of size $1° \times 1°$ (approximately 110 km $\times$ 110 km), and the seismicity parameters were evaluated at the center of each of these grid cells. The evaluation of these values was done based on the magnitude of completeness (M_c) of the catalog (Reasenberg, 1985). The value of M_c was calculated at the center of the grid points by considering the events within a radius of 300 km.

M_c will have spatial and temporal variation, and it will decrease with time, mainly because of the increase in the number of seismographs in the region. One of the most widely used methods to evaluate M_c is based on a power law fit for the FMD, which was suggested by Wiemer and Wyss (2000). In this method, a series of synthetic magnitude distributions are developed for each magnitude interval using a maximum likelihood estimate.

The M_c value is calculated at the magnitude where the goodness of fit is greater than 90%. The goodness of fit can be tested for 95% also, but this level is rarely obtained for real earthquake catalogs (Woessner and Wiemer, 2005). This analysis was done using ZMAP software (Wiemer, 2001).

The correct estimate of the *a* and *b* values depends critically on the completeness of the sample under investigation. The FMD deviates from a linear power law fit increasingly for smaller magnitudes due to the fact that the recording network is only capable of recording a fraction of all events for magnitudes smaller than the magnitude of completeness, M_c. If M_c is raised to large values, the uncertainty in the *b* value estimate increases strongly. The situation is complicated by the fact that M_c varies as a function of space and time throughout all earthquake catalogs, and hence estimating the correct M_c while maximizing the available number of earthquakes becomes difficult. The method suggested by Wiemer and Wyss (2000) is employed to estimate M_c in the present study, as elaborated in subsequent section. The variation of M_c with space is shown in Figure 4.1.

4.2.2   Estimation of *a* and *b* values

The seismic activity of a region is given by the Gutenberg and Richter (1944) earthquake recurrence law. The recurrence rate given by this law is:

$$\log_{10} N = a - bM \tag{4.1}$$

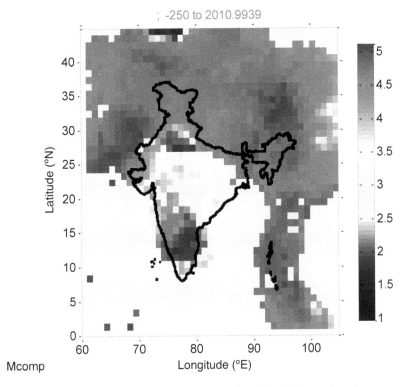

Figure 4.1 Spatial variation of magnitude of completeness from the declustered catalog

where N is the total number of earthquakes with magnitude M and above that will occur in a year and a and b are the seismicity parameters of the region. Seismicity parameters can be evaluated using the maximum likelihood estimation technique (Utsu, 1965; Aki, 1965).

For evaluating the seismicity parameter, the completeness of the catalog has to be analyzed, and the data in the complete part of the catalog need to be used for the analysis. The magnitude of completeness is the lowest magnitude above which the earthquake recording is assumed to be complete (Rydelek and Sacks, 1989). For seismicity studies, the evaluation of the magnitude of completeness (M_c) is very important. Below this magnitude, a fraction of events is missed by the network because they are either too small to be recorded by enough stations, or because they are below the magnitude of interest, or because they are mixed with the coda of a larger event, and therefore they passed undetected. M_c will vary spatially and temporally, and it will decrease with time mainly because of the increase in the number of seismographs in the region. The b value can be evaluated using the maximum likelihood estimate:

$$b = \frac{\log_{10}(e)}{\left[m_{\text{mean}} - \left(M_c - \frac{\Delta m_{\text{bin}}}{2} \right) \right]} \qquad (4.2)$$

where m_{mean} is the mean magnitude of the sample, M_c is the magnitude of completeness, and Δm_{bin} is the magnitude bin size. One of the most widely used methods to evaluate M_c is based on a power law fit for the FMD, which was suggested by Wiemer and Wyss (2000). In this method, a series of synthetic magnitude distributions are developed for each magnitude interval using a maximum likelihood estimate. The a and b values obtained in the synthetic distribution are compared with the observed distribution, and the goodness of fit is calculated. For calculating the goodness of fit, the absolute difference (R) between the observed and synthetic distribution has to be calculated:

$$R(a,b,M_i) = 100 - \left(\frac{\sum\limits_{m_i}^{m_{max}} \text{mod}(B_i - S_i)}{\sum\limits_{i} B_i} *100 \right) \tag{4.3}$$

where B_i and S_i are the observed and predicted cumulative number of events in each magnitude bin. The uncertainties involved in evaluating b values can be calculated using the bootstrap method (Chee nick, 1999).

The evaluation of the b values in the present case was done based on the maximum likelihood method (Aki, 1965). For this calculation, only those earthquake events that were greater than the magnitude of completeness M_c for each grid point were considered. To obtain better estimates of b values, the values were evaluated for those grid points that had at least 50 events with magnitude equal to or greater than M_c. This criterion is essential for getting a good statistical analysis (Utsu, 1999). The uncertainties involved in evaluating the b values were calculated using the bootstrap method with 100 bootstraps (Chernick, 1999). The spatial and temporal variations of seismic activity across the country were investigated based on the seismic tool ZMAP (Wiemer, 2001).

The b value at each grid point was estimated considering the events within a radius of 300 km from the center of the grid. The spatial variation of b values across the study area obtained from clustered and declustered catalogs is shown in Figures 4.2 and 4.3, respectively. The declustered catalog gives lower b values in most of the regions (Koyna region, north and northeast India, and the Andaman and Nicobar Islands). The reason for a significant decrease in seismic b values in the declustered catalog compared to the raw catalog is related to the larger proportion of foreshocks and aftershocks in the raw catalog. The proportion of foreshocks and aftershocks in the earthquake catalog is inversely correlated with earthquake magnitude. It means that a larger proportion of dependent events in the earthquake catalog are related to lower magnitude events. The inclusion of dependent events in the catalog affects the relative abundance of low and high magnitude earthquakes. Thus, greater inclusion of dependent events leads to higher b values and higher activity rate, as is evident from Figures 4.2 and 4.3. Hence, the seismicity parameters obtained from the declustered catalog are valid as they follow a Poisson distribution.

The spatial variation of b values using the declustered catalog was studied by considering both constant radius and constant numbers of earthquakes. The b value map obtained using a constant radius of 300 km is given in Figure 4.3 and that obtained using the nearest 200 events is shown in Figure 4.4. Both approaches are equally valid, and from the comparison

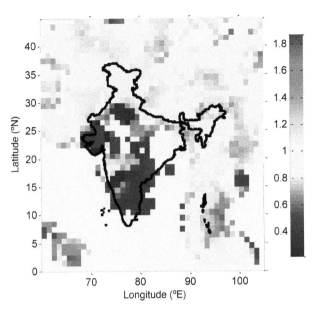

Figure 4.2 Spatial variation of *b* values from the clustered catalog considering the events within a radius of 300 km from the center of each grid point

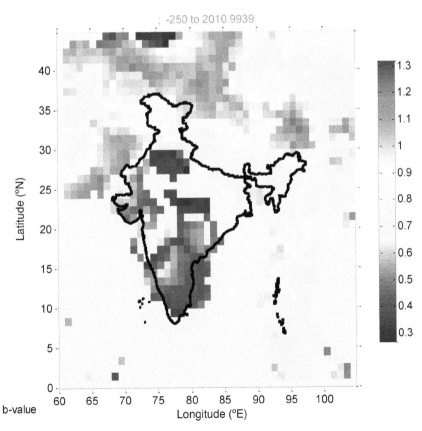

Figure 4.3 Spatial variation of *b* values from the declustered catalog considering the events within a radius of 300 km from the center of each grid point

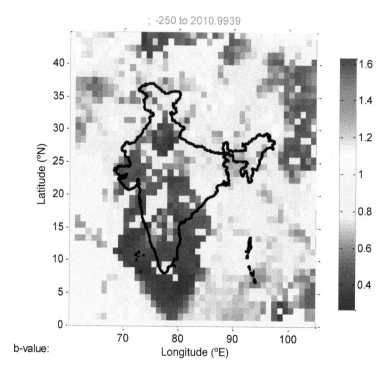

; -250 to 2010.9939

b-value:

Figure 4.4 Spatial variation of *b* values from the declustered catalog considering the nearest 200 events from a grid point

Table 4.1 Comparison of *b* values obtained from different methodologies

Longitude (°)	Latitude (°)	b value with events within 300 km	b value with nearest 200 events
80	15	0.58	0.53
80	20	0.50	0.50
90	25	0.65	0.67
80	35	0.89	0.85

of both the results it is seen that seismicity parameters are almost independent of the choice of the sampling method (Table 4.1). By sampling a constant number of events at each node, the sample size, and hence uncertainty, is approximately constant, and the best spatial resolution possible at each node is achieved (Wiemer and Wyss, 2002). In this case, the radii of sampling volumes, or resolution, is inversely proportional to the local density of earthquakes, and consequently variable across a region. When using constant radii for sampling, the resolution does not vary spatially, but the sample size, and hence the uncertainty, varies. The constant radius method seems to be more valid as it characterizes the seismicity of a region with respect to a defined space limit.

The *b* value in the region varies from 0.5 to 1.5, and for the majority of the study area the value is around 1. The *a* value for the study area varies from a lower value of 3 to a higher value of 10

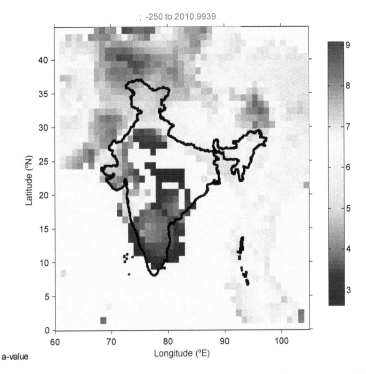

Figure 4.5 Spatial variation of *a* values from the declustered catalog considering events within a radius of 300 km from the center of each grid point

(Fig. 4.5). The *a* and *b* values were not evaluated for some of the regions (shown as void regions in Figs. 4.2, 4.3, 4.4, and 4.5), as those cells did not have an adequate number of earthquake events (50) with magnitude equal to or greater than M_c. Higher *b* values are observed in north and northeast India and in the Andaman and Nicobar Islands. The lower *b* values obtained in shield regions implies that the energy released in these regions is mostly from large magnitude events. The *b* value of northeast India and the Andaman and Nicobar Island region is around unity, which implies that the energy released is compatible for both smaller and larger events.

4.3   DELINEATION OF SEISMIC SOURCE ZONES

It is important to identify and delineate regional seismic source zones when conducting a seismic hazard analysis. The catalog prepared is heterogeneous with space, and the catalog cannot be used in a meaningful way as a whole to estimate the seismicity of the study area. The zones of different seismicity characteristics should be defined based on space limits, and a separate subset of the catalog should be used for each of these zones.

The Indian subcontinent has a complex seismotectonic setup. The country is spread over a vast region, and the earthquake pattern obtained from the past earthquake data (Fig. 2.1) indicates nonuniformity in seismic activity. The possible seismic sources in the study area were identified by Gupta (2006) for all of India and by Seeber et al. (1999) for south India. Gupta (2006) divided the Indian region into 81 zones based on the trend of tectonic features,

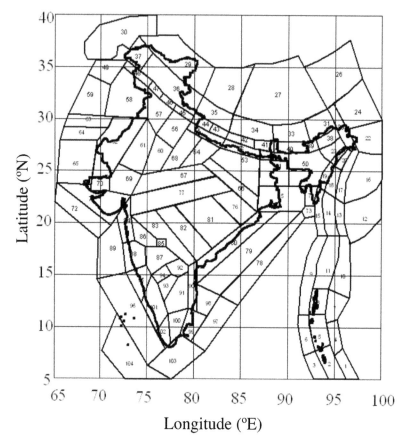

Figure 4.6 Identified regional seismic source zones in and around India

Source: Kolathayar and Sitharam (2012).

predominant source mechanism solutions, and the epicentral distribution of past earthquakes. Seeber et al. (1999) have identified nine broad seismic zones in south India based on tectonic features and past earthquakes in the peninsular shield. In the present study, Gupta (2006) and Seeber et al.(1999) were taken as the main reference, and 104 regional seismic source zones were carefully identified based on the pattern of seismic event distribution and spatial variation of seismicity parameters. The identified seismic source zones are shown in Figure 4.6. These zones allow for local variability in seismicity characteristics, such as changes in b, a, M_{max}, etc.

4.4   EVALUATION OF M_{MAX} FOR DIFFERENT SEISMIC SOURCE ZONES

Another important parameter in the seismic hazard evaluation is the maximum possible earthquake magnitude (M_{max}). This defines the upper limit of earthquake magnitude for a given seismic source zone or an entire region. Researchers have tried to evaluate M_{max} as a function of fault type and the seismic source rupture area (Wells and Coppersmith, 1994; Anderson et al., 1996, etc.). Attempts have been made to evaluate M_{max} based on the strain rate or the seismic moment release rate (Stein and Hanks, 1998; Field et al., 1999).

The deterministic estimates of M_{max} consist of many uncertainties, and they are subjective. Attempts to develop probabilistic methods to evaluate M_{max} began in the early 1960s. One of the most commonly used probabilistic approaches for assessing M_{max} was proposed by Kijko and Sellevoll (1989). This method was developed based on the assumption of the doubly truncated Gutenberg-Richter relation:

$$F_M(M) = \begin{cases} 0 & for\ M < M_{min} \\ \dfrac{1-\exp[-\beta(M-M_{min})]}{1-\exp[-\beta(M_{max}-M_{min})]} & for\ M_{min} \leq M \leq M_{max} \\ 1 & for\ M > M_{min} \end{cases} \tag{4.4}$$

where $\beta = b\ ln(10)$. The value of M_{max} can be evaluated from the Equation 4.4 based on two different approaches: Cramer's approximation (2016) and Bayesian methods. The solution based on Cramer's approximation is:

$$M_{max} = M_{max}^{obs} + \frac{E_1(n_2) - E_1(n_1)}{\beta \exp(-n_2)} + M_{min} \exp(-n) \tag{4.5}$$

where M_{max}^{obs} is the largest observed magnitude; n is the number of earthquakes with magnitude equal to or greater than M_{min}; n_1 is $n/\{1 - \exp[-\beta(M_{max} - M_{min})]\}$; n_2 is $n_1\{1 - \exp[-\beta(M_{max} - M_{min})]\}$; and $E_1(z)$ ($z = n_1$ or n_2) can be defined as $E_1(z) = \int_z^{\infty} \exp(-\xi)/\xi d\xi$. This expression can be conveniently approximated as $E_1(z) = \dfrac{z^2 + a_1 z + a_2}{z(z^2 + b_1 z + b_2)} \exp(-z)$; $a_1 = 2.334733$, $a_2 = 0.250621$, $b_1 = 3.330657$, and $b_2 = 1.681534$ (Abramowitz and Stegun, 1970). Equation 4.5 has to be solved iteratively to obtain the value of M_{max}. The solution can also be done based on a Bayesian approach (Kijko, 2004). The solution obtained from this approach is:

$$M_{max} = M_{max}^{obs} + \frac{\delta^{1/q+2} \exp(nr^q/(1-r))}{\beta} \left[\Gamma(-1/q, \delta \cdot r^q) - \Gamma(-1/q, \delta) \right] \tag{4.8}$$

where $r = p/(p + M_{max} - M_{min})$, $\delta = nC_\beta$; $p = \bar{\beta}/(\sigma_\beta)^2$; $q = (\bar{\beta}/\sigma_\beta)^2$; $\bar{\beta}$ is the mean value of β; σ_β is the standard deviation of β; $C_\beta = \{1 - [p/(p + M_{max} - M_{min})]^q\}^{-1}$; and $\Gamma(,)$ is the complimentary incomplete gamma function.

The values of M_{max} for each source zone were calculated based on the above methods.

4.5   ESTIMATION OF SEISMICITY PARAMETERS FOR SOURCE ZONES

The correct estimation of a and b values depends critically on the completeness of the sample under investigation. The FMD deviates from a linear power law fit increasingly for smaller magnitudes, due to the fact that the recording network is only capable of recording a fraction of all events for magnitudes smaller than the magnitude of completeness, M_c. If M_c is raised to large values, the uncertainty in the b value estimate increases strongly. The situation is complicated by the fact that M_c varies as a function of space and time throughout all earthquake catalogs, hence estimating the correct M_c, while maximizing the available number of earthquakes, becomes difficult. The method suggested by Wiemer and Wyss (2000) is adopted to estimate M_c in the present study. M_c was estimated for each source zone, and the a and b values were estimated considering all the events in that zone with M_W greater than or equal to M_c. The FMD plot for zone 1 is shown in Figure 4.7. The FMD plots for all the zones are given in Appendix B.

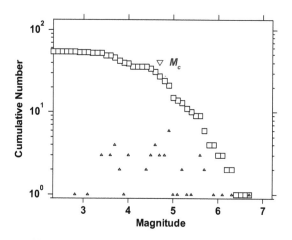

Maximum Likelihood Estimate, Uncertainties by bootstrapping
b-value = 0.82 +/- 0.27, a value = 5.38, a value (annual) = 3.58
Magnitude of Completeness = 4.7 +/- 0.42

Figure 4.7 Frequency magnitude distribution for source zone I

The seismicity parameters estimated for various source zones are given in Table 4.2. The b values for various regions vary from a lower value of 0.5 to a higher value of 1.5. The a values for different zones vary from a lower value of 2 to a higher value of 10. The a and b values were not evaluated for three source zones where the seismic activity is very low (no events reported with $M_W \geq 4$), and those zones are considered to be inactive. Higher a values are observed in north and northeast India and in the Andaman and Nicobar Islands. The lower a values obtained in shield regions implies that the average number of earthquakes per year in these regions is very low.

Table 4.2 Seismicity parameters of regional seismic source zones identified in India

Source zone	No. of events in the zone	b	Standard deviation in b	a	M_{max}	Magnitude of completeness (M_c)
1	54	0.82	0.27	5.38	6.70	4.70
2	310	0.69	0.18	5.65	7.10	4.90
3	93	0.72	0.16	5.20	7.10	4.70
4	38	0.51	0.07	3.55	6.90	4.20
5	378	0.97	0.14	7.05	6.70	4.90
6	64	1.34	0.35	7.90	6.20	4.80
7	173	0.73	0.11	5.53	6.50	4.70
8	236	0.85	0.13	6.22	8.60	4.90
9	39	1.13	0.18	6.92	6.20	4.90
10	137	0.79	0.16	5.62	6.90	4.70
11	134	1.13	0.15	7.23	5.80	4.70
12	377	0.79	0.09	5.94	7.60	4.60
13	58	1.68	1.21	10.10	6.80	4.90
14	325	0.86	0.14	6.24	6.90	4.70
15	140	0.98	0.25	6.42	6.50	4.60
16	629	0.76	0.07	5.90	7.50	4.60
17	300	0.71	0.07	5.57	8.20	4.60
18	479	0.79	0.08	6.09	7.40	4.70
19	168	0.75	0.17	5.44	7.30	4.60
20	138	1.39	0.51	8.85	6.20	5.10
21	39	0.68	0.25	4.47	6.40	4.40
22	219	0.90	0.28	6.20	6.70	4.50
23	41	0.96	0.22	5.96	6.60	4.70
24	286	1.11	0.47	7.48	6.80	4.70
25	215	1.01	0.26	6.99	8.60	4.90
26	354	1.02	0.32	7.25	6.50	4.90
27	980	0.85	0.07	6.73	8.00	4.70
28	415	0.85	0.14	6.22	6.80	4.60
29	1510	1.01	0.07	7.52	7.40	4.70
30	8063	0.90	0.08	8.04	8.10	4.70
31	55	0.91	0.44	5.58	5.80	4.30
32	25	0.76	0.00	4.67	6.30	4.80
33	82	0.74	0.20	5.14	6.20	4.50
34	238	0.84	0.20	6.02	6.50	4.60
35	227	0.79	0.17	5.59	6.60	4.50
36	1016	0.87	0.07	6.70	7.10	4.60
37	967	0.93	0.08	6.97	7.20	4.70
38	68	0.97	0.29	6.13	6.20	4.60
39	87	0.90	0.14	6.16	6.80	4.80
40	49	0.82	0.56	5.64	6.60	5.00

(Continued)

Table 4.2 (Continued)

Source zone	No. of events in the zone	b	Standard deviation in b	a	M_{max}	Magnitude of completeness (M_c)
41	115	1.34	0.53	8.63	8.00	5.20
42	126	0.80	0.16	5.64	7.10	4.70
43	65	1.22	0.61	7.37	5.70	4.70
44	318	0.83	0.10	6.07	7.70	4.70
45	182	1.00	0.36	6.54	7.10	4.50
46	160	0.69	0.39	4.70	6.20	3.70
47	306	0.65	0.25	4.90	7.80	4.10
48	317	0.67	0.14	5.18	7.60	4.50
49	749	0.80	0.07	6.19	7.70	4.60
50	379	0.72	0.09	5.76	8.10	4.70
51	88	0.76	0.33	5.41	7.00	4.80
52	33	0.62	0.14	4.21	8.10	4.70
53	54	0.95	0.43	5.89	5.90	4.60
54	62	0.81	0.34	5.33	6.20	4.50
55	67	0.93	0.42	5.84	7.10	4.60
56	635	0.40	0.02	3.45	7.50	1.90
57	231	0.97	0.08	6.33	6.10	4.50
58	977	0.86	0.08	6.70	7.40	4.60
59	559	1.07	0.19	7.57	7.30	4.90
60	54	0.55	0.08	3.28	5.20	3.50
61	132	0.48	0.33	3.53	5.60	2.80
62	127	0.70	0.47	5.05	6.60	4.20
63	555	1.01	0.17	7.29	7.50	4.90
64	220	0.97	0.27	6.72	6.40	4.70
65	397	1.00	0.17	7.15	7.60	4.90
66	17	0.45	0.08	2.63	5.00	3.10
67	24	0.89	0.00	5.49	6.20	5.20
68	17	0.63	0.17	3.42	5.40	3.80
69	40	0.72	0.46	4.77	6.50	4.50
70	205	0.43	0.08	3.53	8.30	3.00
71	81	0.63	0.50	4.44	6.30	3.90
72	34	1.63	0.80	9.40	5.70	4.80
73	14	0.66	0.33	3.95	5.60	4.10
74	52	0.74	0.14	5.02	6.90	4.70
75	67	0.75	0.31	5.11	6.50	4.50
76	24	1.17	0.00	6.50	6.30	4.60
77	113	0.49	0.30	3.75	6.60	3.60
78	12	0.58	0.00	3.29	6.60	4.80
79	37	0.97	0.15	5.90	6.10	4.70
80	84	1.27	0.88	7.15	5.70	4.00

Source zone	No. of events in the zone	b	Standard deviation in b	a	M_{max}	Magnitude of completeness (M_c)
81	43	0.59	0.19	3.26	5.90	3.00
82	87	0.54	0.07	3.40	6.20	3.00
83	71	0.50	0.12	3.19	5.70	3.20
84	77	0.56	0.21	3.81	7.00	3.80
85	35	0.49	0.09	2.99	6.30	3.60
86	36	1.24	0.75	6.67	5.30	4.00
87	31	0.54	0.00	2.97	5.20	3.50
88	497	0.67	0.38	5.10	6.50	3.40
89	15	0.40	0.00	2.05	6.20	3.90
90	124	0.82	0.64	5.29	6.20	3.70
91	647	0.44	0.02	3.56	5.70	2.00
92	57	1.08	0.22	3.89	4.00	2.20
93	132	0.55	0.07	2.70	4.80	1.10
94	280	1.58	0.19	5.80	6.20	2.20
95	14				3.90	
96	14	0.41	0.11	2.44	5.20	3.40
97	16	0.45	0.13	2.44	6.30	3.20
98	19	—	—	—	3.50	—
99	6	—	—	—	2.60	—
100	14	0.51	0.22	2.34	4.90	2.40
101	207	0.57	0.06	3.28	6.00	2.10
102	30	0.73	0.23	3.97	5.30	3.80
103	14	0.54	0.00	2.99	6.00	4.00
104	12	0.47	0.26	3.05	5.80	4.00

4.6 SUMMARY

A quantitative study of the spatial distribution of the seismicity rate across India and its vicinity was presented in this chapter. The lower b values obtained in shield regions implies that the energy released in these regions is mostly from large magnitude events. The b values of northeast India and the Andaman and Nicobar Island region are around unity, which implies that the energy released is compatible for both smaller and larger events.

The effect of aftershocks in the seismicity parameter was also studied. Maximum likelihood estimations of the b value from the raw and declustered earthquake catalogs show significant changes, leading to a larger proportion of low magnitude events as foreshocks and aftershocks. The inclusion of dependent events in the catalog affects the relative abundance of low and high magnitude earthquakes. Thus, greater inclusion of dependent events leads to higher b values and higher activity rates. Hence, the seismicity parameters obtained from the declustered catalog are valid as they tend to follow a Poisson distribution. M_{max} does not change significantly because it depends on the largest observed magnitude rather than the inclusion of dependent events (foreshocks and aftershocks). The spatial variation of the seismicity parameters can be used as a base to identify regions of similar characteristics and to delineate regional seismic source zones.

Regions of similar seismicity characteristics were identified, and 104 regional seismic source zones were delineated, which are an inevitable for a probabilistic seismic hazard analysis. Separate subsets of the catalog were created for each of these zones, and seismicity analysis was performed for each zone after estimating the cutoff magnitude. The coordinates of these source zones and the estimated seismicity parameters a, b, and M_{max} can be a direct input into the probabilistic seismic hazard analysis (PSHA). The earthquake catalog, the coordinates of source zones, the list of events in each zone, and the FMD plots for all source zones are given in Appendix B.

Chapter 5

Probabilistic seismic
hazard assessment

5.1 INTRODUCTION

Probabilistic seismic hazard analysis (PSHA) is a well-established methodology, and was initially developed by Cornell (1968). It is a well-known fact that many uncertainties are involved in earthquake location and size, which makes the problem complex. PSHA provides a framework in which these uncertainties can be identified, quantified, and combined in a rational manner to give a complete picture of the seismic hazard (Kramer, 1996). Deterministic seismic hazard analysis (DSHA) considers just one (or sometimes a few) maximum magnitude–distance scenario, whereas PSHA considers contributions from all the earthquakes occurring at a source. PSHA also considers the effect of earthquake occurrence at any location in the fault. Thus, it considers the uncertainties in (1) the location of earthquake occurrence, (2) the magnitude of the earthquake, (3) the source-to-site distances, and (4) attenuation relations. The most recent knowledge of seismic activity in the region has to be used to evaluate the hazard, incorporating uncertainty associated with different modeling parameters, as well as spatial and temporal uncertainties. Peak ground acceleration (PGA) estimated from PSHA handles immeasurable uncertainties and is restricted to the design of noncritical construction and planning. This chapter describes in detail the methodology adopted for PSHA to evaluate PGA and spectral acceleration (Sa) values at the rock level for the study area. The hazard analysis will be described in detail with emphasis on the unique approach followed in the present study where different methodologies were adopted in modeling the sources and other parameters.

5.2 METHODOLOGY

The evaluation of seismic hazard involves quantification of uncertainties in earthquake magnitude, location, recurrence rate, and the attenuation characteristics of seismic waves. The adoption of probabilistic approaches in seismic hazard analysis provides a framework for identifying, quantifying, and combining these uncertainties in a rational way (Kramer, 1996). As noted, PSHA was initially developed by Cornell (1968). Many researchers have adopted this methodology for evaluating hazards, and recently this method has been adopted in India by Iyengar and Ghosh (2004), Raghu Kanth and Iyengar (2006), Anbazhagan et al. (2009), and Mahajan et al. (2009) for PSHA studies of Delhi, Mumbai, Bangalore, and the northwest Himalayan regions, respectively.

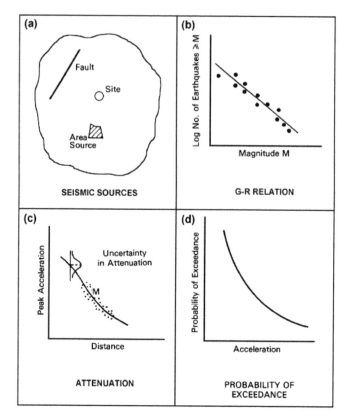

Figure 5.1 Various steps involved in PSHA

Source: Kramer (1996).

The important steps involved in PSHA (Fig. 5.1) are:

1. Identification and characterization of earthquake sources.
2. Characterization of the earthquake recurrence rate.
3. Evaluation of ground motion using the attenuation relationships.
4. Determination of the mean annual rate of exceedance of the ground motion param-
    eter by considering the uncertainties in earthquake location, size, and attenuation
    relation.

The two types of uncertainties encountered in seismic hazard analysis are aleatory and epistemic. The uncertainty of the data is the cause of aleatory variability, and it accounts for
the randomness associated with the results obtained from a particular model. The incomplete knowledge in the predictive models causes the epistemic uncertainty (modeling uncertainty). The aleatory variability of the attenuation relations is taken into account in the PSHA
by considering the standard deviation of the model error. The epistemic uncertainty is considered by multiple models for the evaluation of seismic hazard and combining them using
a logic tree.

5.3 EVALUATION OF GROUND MOTION

The seismic hazard evaluation of area region based on a state-of-the-art PSHA study can be performed using the classical Cornell-McGuire approach (Cornell, 1968). The hazard curves obtained from PSHA show the variation of peak ground acceleration (PGA) or spectral acceleration (Sa) against the mean annual rate of exceedance. For calculating the seismic hazard values, the entire study area need to be divided into grids of required size (depending on the scale of hazard estimation), and the hazard values at the center of each grid cell can be calculated by considering all the seismic sources within a radius of 300 km. The occurrence of an earthquake in a seismic source is assumed to follow a Poisson distribution. That the probability of ground motion parameter, Z, at a given site, will exceed a specified level, z, during a specified time, T is represented by the expression:

$$P(Z > z) = 1 - e^{-v(z)T} \leq v(z)T \tag{5.1}$$

where $v(z)$ is the mean annual rate of exceedance of ground motion parameter, Z, with respect to z. The function $v(z)$ incorporates the uncertainty in time, size, and location of future earthquakes and uncertainty in the ground motion they produce at the site. It is given by:

$$v(z) = \sum_{n=1}^{N} N_n(m_0) \int_{m=m_0}^{m_{max}} f_n(m) \left[\int_{r=0}^{\infty} f_n(r \mid m) P(Z > z \mid m, r) dr \right] dm \tag{5.2}$$

where N_n (m_0) is the frequency of earthquakes on a seismic source n, having a magnitude equal to or greater than a minimum magnitude m_0 (in this study it is taken as 4.0); $f_n(m)$ is the probability density function for a minimum magnitude of m_0 and a maximum magnitude of m_{max}; $f_n(r \mid m)$ is the conditional probability density function (probability of occurrence of an earthquake of magnitude m at a distance r from the site for a seismic source n); $P(Z > z \mid m, r)$ is the probability at which the ground motion parameter Z exceeds a predefined value of z, when an earthquake of magnitude m occurs at a distance of r from the site. The integral in Equation 5.2 can be replaced by summation and the density functions $f_n(m)$, and $f_n(r \mid m)$ can be replaced by discrete mass functions. The resulting expression for $v(z)$ is given by:

$$v(z) = \sum_{n=1}^{N} \sum_{m_i=m_0}^{m_i=m_{max}} \lambda_n(m_i) \left[\sum_{r_j=r_{min}}^{r_j=r_{max}} P_n(R = r_j \mid m_i) P(Z > z \mid m_i, r_j) \right] \tag{5.3}$$

where $\lambda_n(m_i)$ is the frequency of occurrence of magnitude m_i at the source n obtained by discretizing the earthquake recurrence relationship for the source n.

5.3.1 Magnitude recurrence rate

The magnitude recurrence model for a seismic source specifies the frequency of earthquake events with various magnitudes per year. The methods adopted for evaluation of the seismicity parameters were discussed in detail in previous chapters. The maximum magnitude

(m_{max}) for each linear source (fault) was estimated based on the maximum reported magnitude at the source plus 0.5, whereas in the case of the smoothed seismic sources the m_{max} values were evaluated for each of the source zones, and these values were used for all the sources within that zone. The values of m_{max} obtained for different source zones are given in Chapter 4.

The recurrence relation for each linear source capable of producing earthquake magnitudes in the range m_0 to m_{max} was calculated using the truncated exponential recurrence model developed by McGuire and Arabas (1990), and it is given by the following expression:

$$N(m) = N_i(m_0) \frac{e^{-\beta(m-m_0)} - e^{-\beta(m_{max}-m_0)}}{1 - e^{-\beta(m_{max}-m_0)}} \quad \text{for } m_0 \leq m \leq m_{max} \tag{5.4}$$

where $\beta = b\ln(10)$ and $N_i(m_0)$ is the weighting factor for a particular fault based on deaggregation.

The evaluation of seismicity parameters was done based on the data from a source zone and not based on the individual sources (faults). For differentiating the fault-level activity rate, it is necessary to find the fault-level recurrence rate. A given fault can produce an earthquake of magnitude m, which will be within the limits of m_0 and m_{max}, based on the earthquake recurrence rate of that fault. However, this recurrence rate of the faults is not known; instead what is available is the recurrence rate of the seismic source zone or the entire region. Hence, from the regional recurrence rate, the fault-level recurrence rates were calculated based on the principle of conservation of seismic activity (Iyengar and Ghosh, 2004; Raghu Kanth and Iyengar, 2006). This was achieved by considering two weighting factors that are based on the fault length and number of earthquake events recorded along the fault. The number of earthquakes that a fault can produce is taken as proportional to its length. The weighting factor for length is:

$$\alpha_i = \frac{L_i}{\sum L_i} \tag{5.5}$$

where L_i is the length of the fault i and $\sum L_i$ is the length of all the faults considered in the source zone or region. The weighting factor based on the number of earthquake events (χ_i) has been taken as the ratio of the past events associated with source i to the total number of events in the region:

$$\chi_i = \frac{\text{Number of earthquakes along the source } i}{\text{Total number of earthquakes in the region}} \tag{5.6}$$

The recurrence relation for fault i has been calculated by averaging both weighting factors:

$$N_i(m_0) = 0.5(\alpha_i + \chi_i)N(m_0) \tag{5.7}$$

A similar procedure was adopted by Iyengar and Ghosh (2004), Raghu Kanth and Iyengar (2006), and Anbazhagan et al. (2009) for the evaluation of recurrence rate at the fault level. The value of m_0 for all the faults was taken as 4, and the value of m_{max} for each of the faults was calculated as explained above. The magnitude range for each fault was divided into small intervals, and the recurrence rate of each of these magnitudes was calculated using the above equations.

5.3.2   Evaluation of hypocentral uncertainty

In PSHA, the probability of occurrence of an earthquake anywhere along the fault is assumed to be the same. The uncertainty involved in the source-to-site distance is described by a probability density function. The shortest and longest distance from each fault to the center of the grid is calculated. The hypocentral distance was calculated by considering a focal depth of 15 km. This focal depth was selected based on the observations of focal depths made by several other researchers (Mandal et al., 1998; Patrol et al., 2006; Kayal, 2007; Anbazhagan et al., 2009). A probable source zone depth of 10 km was considered by Bhatia et al. (1999) in an exercise to develop a seismic hazard map of the shield region of India in GSHAP. The conditional probability distribution function of the hypocentral distance R for an earthquake of magnitude m is assumed to be uniformly distributed along a fault. It is given by Kiureghian and Ang (1977) as:

$$P(R < r \mid M = m) = 0 \text{ for } R < (D^2 + L_0^2)^{1/2} \tag{5.8}$$

$$P(R < r \mid M = m) = \frac{(r^2 - D^2)^{1/2} - L_0}{L - X(m)} \text{ for } (D^2 + L_0^2) \leq R < \{D^2 + [L + L_0 - X(m)]^2\}^{1/2} \tag{5.9}$$

$$P(R < r \mid M = m) = 1 \text{ for } R > \{D^2 + [L + L_0 - X(m)]^2\}^{1/2} \tag{5.10}$$

where $X(m)$ is the rupture length in kilometers, for an event of magnitude m, estimated using the Wells and Coppersmith (1994) equation:

$$X(m) = \min [10^{(-2.44+0.59m)}, \text{ fault length}] \tag{5.11}$$

The notations used in the above equations are shown in Figure 5.2.

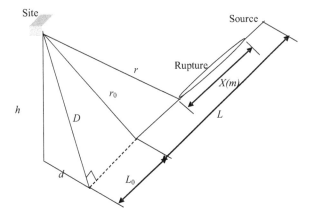

Figure 5.2 Fault rupture model

Source: Wells and Coppersmith (1994).

5.3.3   Uncertainty in the attenuation relationship

The aleatory variability in a GMPE is characterized by the standard deviation that is assumed to follow a log-normal distribution (Frank et al., 2005). In PSHA, this influence is taken into account by integrating across the log-normal scatter in the prediction equations as part of calculating the exceedance frequency of different levels of ground motion. The normal cumulative distribution function has a value that is most efficiently expressed in terms of a standard normal variable (z),which can be computed for any random variable using the following transformation (Kramer, 1996):

$$z = \frac{\ln PHA - \overline{\ln PHA}}{\sigma_{\ln PHA}} \tag{5.12}$$

where PHA (Peak Horizontal Acceleration) is the various targeted peak acceleration levels that will be exceeded, $\overline{\ln PHA}$ is the value calculated using the attenuation relationship, and $\sigma_{\ln PHA}$ is the uncertainty in the attenuation relation expressed by the standard deviation.

5.4   PSHA OF INDIAN SUBCONTINENT – A CASE STUDY

Seismic hazard evaluation of the Indian landmass based on state-of-the-art PSHA was performed using the classical Cornell–McGuire approach with different source models and attenuation relations. The most recent knowledge of seismic activity in the region was used to evaluate the hazard, incorporating uncertainty associated with different modeling parameters, as well as spatial and temporal uncertainties. PSHA was performed with up-to-date data and the best possible scientific interpretation using an appropriate instrument, the logic tree, to explicitly account for epistemic uncertainty by considering alternative models (i.e., source models, the maximum magnitude in hazard computations, and ground motion attenuation relationships).

5.4.1   Logic tree structure

The models used for seismic hazard assessment have many uncertainties, and this makes the selection of a seismic hazard model difficult. The use of the logic tree approach allows characterization of epistemic uncertainties in various models by including alternative models in the analysis (Budnitz et al., 1997; Stepp et al., 2001; Bommer et al., 2005). The logic tree consists of a series of nodes and branches, and these branches denote different models (hypotheses). A subjective weighting, based on engineering judgment, can be given to each of these branches depending on the likelihood of being correct.

The weight for all the branches at a particular node should be equal to unity. The weight of the terminal branch of the logic tree can be obtained by multiplying the weights of all the branches leading to it. The present study considers three types of source models and three different attenuation relations for various tectonic provinces in the study area. The source models employed are linear sources, gridded seismicity models, and areal sources. Two different maximum magnitudes have been considered for linear sources: the maximum historical magnitude and the previous increased by 0.5 units. This is the criterion followed and defined by Gupta (2002) for assigning the maximum magnitude to a linear source. Similar methods were adopted by Iyengar and Ghosh (2004) and Raghu Kanth and Iyengar (2006) for the seismic hazard analysis of Delhi and Mumbai, respectively. In a recent work, Bozzoni et al. (2011) used the same criteria to assign the maximum magnitude to a regional source zone. The usual trend followed is to use the M_{max} estimated for the source zone to be assigned to all linear

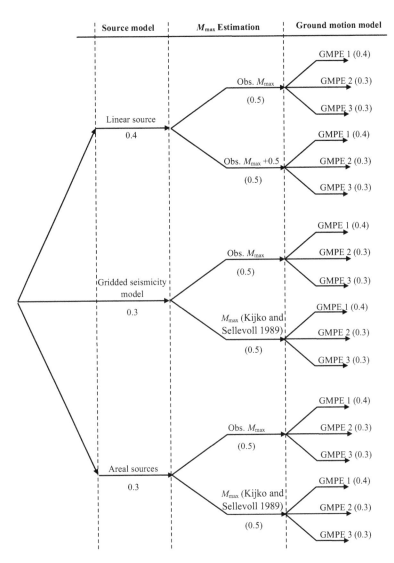

Figure 5.3 Parameters and weighting factors adopted in the logic tree

Source: Sitharam et al. (2015).

sources in that zone. Here we have followed a finer approach, identifying M_{max} associated with each fault separately. The attenuation relations used for each tectonic province are the same as described in Chapter 3 (Table 3.1). Many uncertainties are associated with these models, and to get a reasonable value of hazard, different models were combined in the logic tree framework. The logic tree branch with the weights assigned to each model is shown in Figure 5.3.

5.4.2 Evaluation of PGA

The seismic hazard analysis of India was done by dividing the entire country into grids of size $0.1° \times 0.1°$ (about 11 km × 11 km). The total number of grid points considered for the analysis was 29,889. For each grid point, the PGA and Sa (for periods of 0.1 s and 1 s) values were evaluated at bedrock level corresponding to a probability of exceedance (PE) of 10% and 2% in 50 years. These exceedance values correspond to return periods of 475 and 2475 years respectively. A 10% probability of exceedance in 50 years corresponds to ground motions having 90% probability of not being exceeded in the next 50 years. The 90% nonexceedance probability and the 50 years is exposure time. It can be described such that these are 475-year return-period ground motions, which is the same as saying that these ground motions have an annual probability of occurrence of 1/475 per year. The return period is the inverse of the annual probability of occurrence (i.e., of getting an exceedance of that ground motion).

When evaluating the seismic hazard using PSHA, the hypocentral and magnitude uncertainties were considered. This was done by deaggregating the hypocentral distance into small intervals of 2 km and the magnitude range (between minimum and maximum magnitude) into small incremental values of 0.2 [a lower value of the hypocentral incremental distance (less than 2 km) and magnitude interval (0.2) did not improve the accuracy of the results significantly and increased the computation time]. For each grid point, all sources within a radius of 300 km were considered for evaluation of PGA and Sa values. For the plate boundary regions, because the earthquake magnitudes were higher, this distance was increased to 500 km.

The spatial variation of PGA values obtained for 10% and 2% PE in 50 years are shown in Figures 5.4 and 5.5. PGA values at the rock level for the 10 most populous cities in India are provided in Table 5.1. It can be seen that the seismic hazard is high along the plate boundary

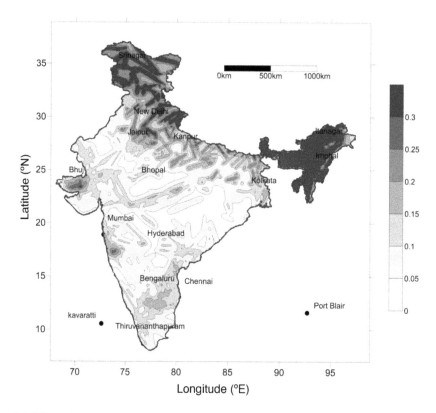

Figure 5.4 PGA values (g) corresponding to a return period of 475 years for 10% probability of exceedance in 50 years

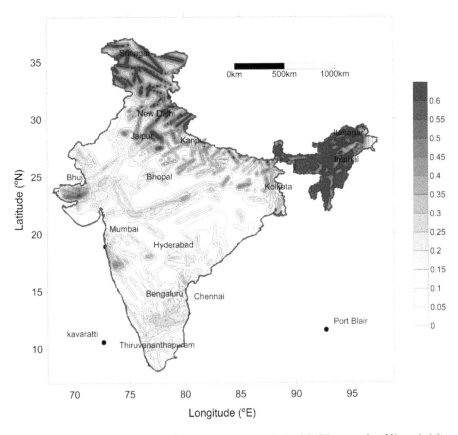

Figure 5.5 PGA values (g) corresponding to a return period of 2475 years for 2% probability of exceedance in 50 years

Table 5.1 PGA values at rock level for the 10 most populous cities in India

City	Location		PGA value (g)	
	Longitude (°E)	Latitude (°N)	For a 475-year return period	For a 2475-year return period
Mumbai	72.82	18.90	0.10	0.19
Delhi	77.20	28.58	0.27	0.51
Bangalore	77.59	12.98	0.13	0.22
Kolkata	88.33	22.53	0.13	0.23
Chennai	80.25	13.07	0.13	0.21
Hyderabad	78.48	17.38	0.01	0.19
Ahmedabad	72.62	23.00	0.10	0.18
Pune	73.87	18.53	0.07	0.13
Kanpur	80.40	26.47	0.11	0.17
Jaipur	75.87	26.92	0.12	0.2

regions, for example, north and northeast India. Along the shield region, the highest hazard is observed in the Bhuj and Koyna regions. For the 475-year return period, the PGA values for plate boundary regions vary from 0.2 g to 0.4 g, whereas for the shield region the values were less than 0.2 g, except for the Kutch region in Gujarat. The Indian seismic code BIS-1893 (2002) divides the country into four zones: II, II, IV, and V. The maximum expected accelerations in each of these zones are 0.1 g, 0.16 g, 0.24 g, and 0.36 g, respectively. The PGA values obtained in this study for northeast India and most parts of Jammu and Kashmir are higher than that specified by BIS-1893 (2002). The spectral acceleration values for the periods 0.1 s and 1 s obtained for 10% and 2% probability of exceedance in 50 years are shown in Figures 5.6 and 5.7. The results obtained by assigning equal weights to all models are given in Appendix C.

The hazard analysis of the Andaman and Nicobar Islands was performed separately, taking into consideration the complex seismotectonic set up of the region. The spatial variation of PGA values in the Andaman and Nicobar Islands obtained for 10% and 2% PE in 50 years is shown in Figure 5.8. The spectral acceleration values for the periods 0.1 s and 1 s obtained for 10% and 2% probability of exceedance in 50 years are shown in Figures 5.9 and 5.10. The value of PGA for the Andaman and Nicobar Islands region for the 475-year return period varies from 0.2 g to 0.4 g, whereasfor a return period of 2475 years the PGA reaches a value as high as 0.6 g.

Use of a logic tree framework to combine different source models is debatable, and hence an attempt was made to estimate the hazard separately for each source model. The spatial variation of PGA values obtained for 10% PE in 50 years using different source models for the Indian landmass are shown in Figures 5.11, 5.12, and 5.13. The linear source model predicts higher hazard compared to the other two source models, and the gridded seismicity model gives the lower value of hazard at a particular grid point.

The PGA values obtained for the 10 most populous cities (megacities and metros) of India for return periods of 475 and 2475 years, using various source models, are given in Table 5.2. It is observed that with the same attenuation relations used, PGA values vary with the use of different source models. From Table 5.2, it can be seen that the linear source model gives PGA values at a higher side compared to other source models. For most of the cities, the gridded seismicity model gives higher values compared to areal sources, except for Ahmedabad, Chennai, and Kanpur. The reason behind this variation is because the gridded seismicity model accounts for local variation of seismicity at a subtler level. In areal sources, it is assumed that the seismic hazard potential is homogeneous within a source zone, but PGA values at two nearby grid points within a zone will vary, as all the sources lying within 500-km radius are taken into consideration at a particular grid point.

The limitation of using linear sources is that there can be hidden faults, especially in the ocean, that have not been identified yet. For the Himalayan region, the values obtained using linear sources can be used confidently, as the linear seismic sources are well quantified by different researchers. For south India, either of the other two source models is recommended, and it is observed that both give comparable results, but are unequal. It is recommended to use the highest PGA value obtained after comparing the results obtained from these three different source models, to be on a conservative side.

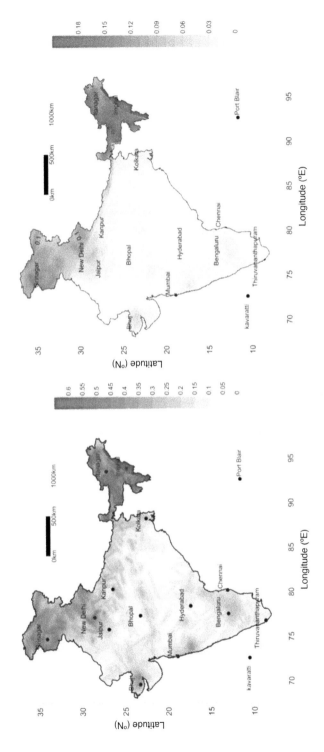

Figure 5.6 Spectral acceleration values (g) at periods 0.1 s and 1 s for 10% probability of exceedance in 50 years

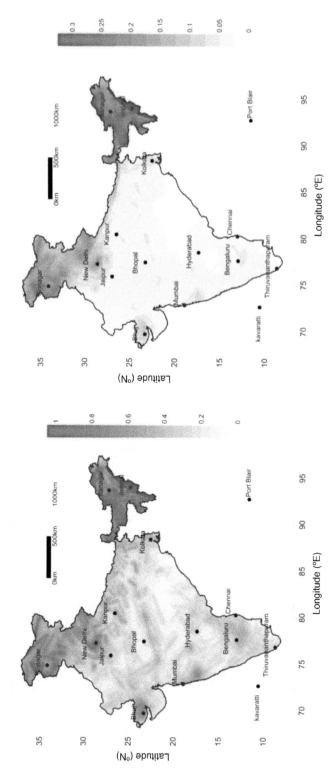

Figure 5.7 Spectral acceleration values (g) at periods 0.1 s and 1 s for 2% probability of exceedance in 50 years

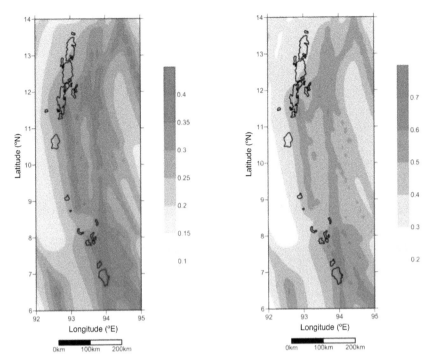

Figure 5.8 Spatial variation of PGA values (g) in the Andaman and Nicobar Islands region corresponding to return periods of (a) 475 years (b) 2475 years

Source: Kolathayar and Sitharam (2012).

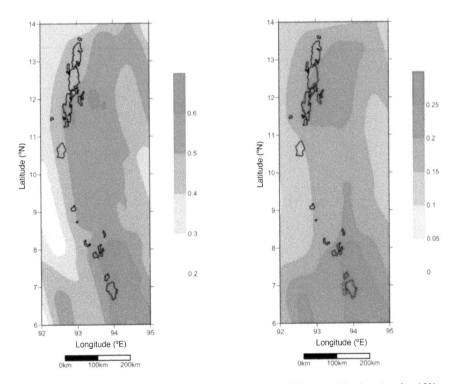

Figure 5.9 Spectral acceleration values (g) in the Andaman and Nicobar Island region for 10% probability of exceedance in 50 years at periods (a) 0.1 s and (b) 1.0 s

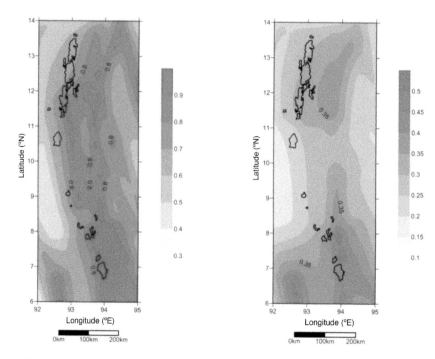

Figure 5.10 Spectral acceleration values (g) in the Andaman and Nicobar Island region for 2% probability of exceedance in 50 years at periods (a) 0.1 s and (b) 1.0 s

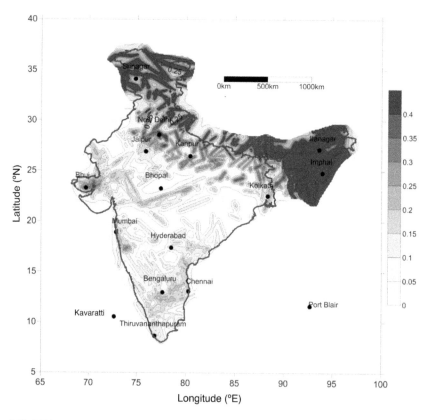

Figure 5.11 PGA values for India corresponding to a return period of 475 years (10% probability of exceedance in 50 years) estimated using linear sources

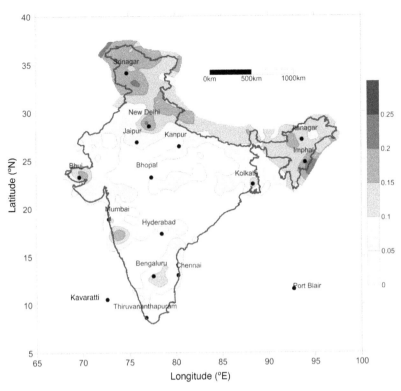

Figure 5.12 PGA values for India corresponding to a return period of 475 years estimated using gridded seismicity source model

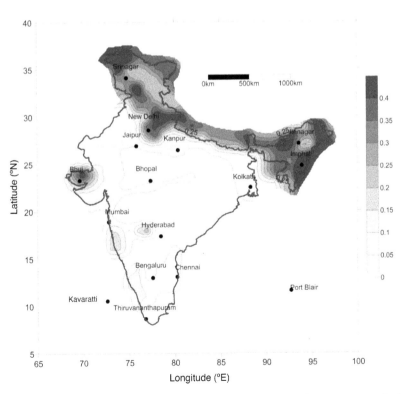

Figure 5.13 PGA values for India corresponding to a return period of 475 years estimated using areal sources

Table 5.2 Comparison of PGA values obtained using different source models, at the rock level for the 10 most populous cities in India

Major cities	PGA value (g)					
	Areal source model		Gridded seismicity model		Linear source model	
	475-year return period	2475-year return period	475-year return period	2475-year return period	475-year return period	2475-year return period
Ahmedabad	0.1	0.20	0.08	0.11	0.12	0.23
Bangalore	0.08	0.13	0.10	0.21	0.12	0.27
Chennai	0.08	0.17	0.06	0.11	0.10	0.21
Delhi	0.16	0.43	0.18	0.36	0.33	0.6
Hyderabad	0.03	0.05	0.055	0.12	0.143	0.24
Jaipur	0.04	0.075	0.074	0.14	0.126	0.21
Kanpur	0.06	0.09	0.05	0.1	0.15	0.21
Kolkata	0.08	0.14	0.12	0.21	0.13	0.24
Mumbai	0.08	0.18	0.09	0.16	0.11	0.2
Pune	0.05	0.10	0.07	0.11	0.08	0.13

The PGA values for plate boundary regions range from 0.3 to 0.5, whereas for the shield region the values are less than 0.25, except for the Kutch region in Gujarat. The results obtained in this study were compared with other studies done for various parts of India by different researchers. The previous PSHA of India was made by Bhatia et al. (1999) as part of Global Seismic Hazard Assessment Program (GSHAP). However, the results obtained in the study of Bhatia et al. (1999) are highly debatable because of the use of a single attenuation relation for the entire country. The comparison of PGA values with those obtained by Seeber et al. (1999) (Maharashtra state); Raghu Kanth and Iyengar (2006) (Mumbai); Iyengar and Ghosh (2004) (Delhi); Jaiswal and Sinha (2007) (peninsular India); Vipin et al. (2009) (south India); Giardini et al. (1999) (Kolkata); Mohanty and Walling (2008) (Kolkata); Menon et al. (2010) (Tamilnadu); Singh (2009) (Ahmedabad); and Iyengar et al. (2010) (Indian land mass) is shown in Table 5.3. DSHA results also have been included for comparison. The results obtained for south India match well with those obtained by Vipin et al. (2009). The results obtained for Tamil Nadu by Menon et al. (2010) and for Maharashtra by Seeber et al. (1999) also match well with the present study. Iyengar et al. (2010) identified 32 seismogenic source zones to estimate the seismicity parameters and linear seismic sources for hazard analysis. They calculated PGA values using attenuation relations that they developed using a ground motion simulation technique. The present study uses 104 regional seismic source zones identified based on the seismic event distribution and fault alignment and three different source models in the hazard analysis. The present study used well-recognized attenuation relations for various seismotectonic provinces of the country that have been developed by multiple researchers. The PGA values reported by Iyengar et al. (2010) for most regions are much lower than those presented in this study. DSHA of the Indian landmass was conducted using point and linear source models, as described in Chapter 3, using the same attenuation relations. The PGA values obtained from DSHA were on the higher side compared to those derived from PSHA.

Table 5.3 Comparison of present PGA values for 475-year return period with those obtained by various researchers for different parts of the country

Location	PGA (g)					
	Present study		Previous studies (PSHA)			
	PSHA	DSHA	Iyengar et al. (2010)	Nath and Thingbaijam (2012)		Other studies
Mumbai	0.10	0.27	0.07	0.16	0.1 Seeber et al. (1999)	0.13 Raghu Kanth and Iyengar (2006)
Delhi	0.27	0.38	0.02	0.24	0.23 Iyengar and Ghosh (2004)	–
Bengaluru	0.13	0.13	0.03	0.11	0.06 to 0.1 Jaiswal and Sinha (2007)	0.13 Sitharam and Vipin (2011)
Kolkata	0.13	0.30	0.08	0.15	0.08 to 0.13 Giardini et al. (1999)	0.22 to 0.24 Mohanty and Walling (2008)
Chennai	0.13	0.1	0.09	0.12	0.09 Menon et al. (2010)	0.08 Jaiswal and Sinha (2007)
Ahmedabad	0.10		0.09	0.11	0.07 Singh (2009)	–

Hazard maps produced for horizontal ground motion at bedrock level (shear wave velocity ≥ 3.6 kms^{-1}) for India were compared with the seismic hazard zoning map by the Indian seismic standards (BIS-1893 part 1, 2002, which is based on intensity and geological data),Bhatia et al. (1999 (India and adjoining areas); Seeber et al. (1999) (Maharashtra state); Jaiswal and Sinha (2007) (peninsular India); Sitharam and Vipin (2011) (south India); and Menon et al. (2010) (Tamilnadu). The PSHA showed that the seismic hazard is moderate in the peninsular shield, but the hazard in north and northeast India is extremely high. The PGA values obtained in this study for northeast India and most parts of Jammu and Kashmir are higher than what is specified by BIS-1893 (2002).

Reliable information on active faults and geodetic measurements of fault movement rates would go a long way in redefining seismogenic sources in India. Strong ground motion records of earthquakes in the region are essential to developing region-specific attenuation equations for various parts of the country. Updated and reliable information on these factors would be instrumental in redefining the seismic hazard of the region.

Chapter 6

Site response and liquefaction analyses

6.1 INTRODUCTION

The modification of seismic waves as they pass from bedrock to the ground surface through the overlying soil is termed the local site effect (Fig. 6.1). It depends on local site conditions, and it is one of the major factors that can increase the destruction caused by an earthquake. This chapter describes in detail the methods adopted for evaluation of peak ground acceleration (PGA) values at the surface level for different site classes. The PGA values for different site classes were evaluated based on the PGA values obtained from DSHA and PSHA.

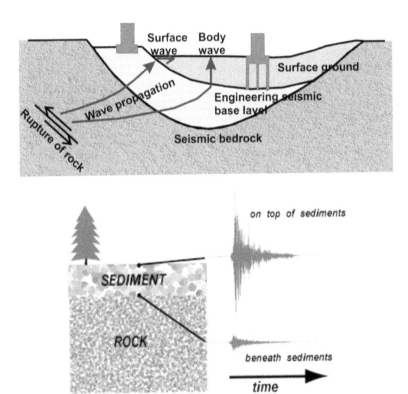

Figure 6.1 (a) Seismic wave propagation from bedrock to surface and (b) amplification of seismic waves passing through the overlying soil

Source: USGS.

Liquefaction is the conversion of formerly stable cohesion less soils to a fluid mass due to increased pore pressure. Liquefaction is an induced hazard in areas that have groundwater near the surface and sandy soil. Soil liquefaction has been observed during earthquakes because of the sudden dynamic earthquake load, which, in turn, increases the pore pressure. The devastating effects of liquefaction have been observed during various earthquakes worldwide, including the 2001 quake that occurred in the Indian state of Gujarat. These instances of liquefaction necessitate the need to evaluate the liquefaction potential of an area. The evaluation of liquefaction potential involves assessment of earthquake loading and soil resistance to liquefaction. Assessment of earthquake loading requires an analysis of the seismotectonic properties of the region, the collection of earthquake details, and evaluation of peak ground acceleration, as has been discussed in previous chapters. Soil resistance depends on factors such as soil properties, age, type of soil deposit, and depth of the groundwater table.

6.2 SITE CLASSIFICATION

When seismic waves travel through the overlying soil, the waves are modified, and this phenomenon is known as site effects. Evaluation of site effects requires classification of the site, which can be based on surface geology, geomorphology, or geotechnical data. One of the most widely used site classification schemes is based on the average shear wave velocity in the top 30 m (V_s^{30}). The National Earthquake Hazard Reduction Program (NEHRP) recommends (The Building Seismic Safety Council (BSSC), 2003) six site classes based on V_s^{30} values. The shear wave velocity ranges for each site class are as follows: site class A ($V_s^{30} > 1.5$ kms^{-1}), site class B (0.76 kms$^{-1} < V_s^{30} \leq 1.5$ kms^{-1}), site class C (0.36 kms$^{-1} < V_s^{30} \leq 0.76$ kms^{-1}), and site class D (0.18 kms$^{-1} < V_s^{30} \leq 0.36$ kms^{-1}). Site class E consists of soils with a soil profile of more than 3 m (10 feet) of clay that has a plasticity index higher than 20 or a water content higher than 40% and $V_s^{30} < 180$ ms^{-1}. Site class F consists of soils such as highly sensitive clays, collapsible weakly cemented soils, etc. Soils of site classes E and F require site-specific evaluations.

6.3 SITE CLASSIFICATION METHODS

Ground acceleration is amplified as seismic waves pass from bedrock to the ground surface. Hence, it is important to characterize the underlying soil to estimate the level of amplification, as different site classes exhibit different amplification characteristics. Accurate knowledge of the geology, geomorphology, and geophysical and geotechnical properties are essential to classify sites. Site classification methods can be broadly divided into the following categories.

6.3.1 Surface geology

Geological maps based on the geologic age of sediments are the most straightforward tool to classify a vast area. This type of classification can also be done with the help of remote sensing maps. The different site classes as per geologic site classification are given in Table 6.1.

Table 6.1 Site classification based on surface geology

Age	Depositional environment	Sediment texture
Holocene	Fan alluvium	Coarse
Pleistocene	Valley alluvium	Fine
	Lacustrine/marine	Mixed
	Aeolian	
	Artificial fill	
Tertiary		
Mesozoic + Igneous		

Surface geology features can be correlated to different site classes given by various codes such as the NEHRP (BSSC, 2003). NEHRP site class B (rock) and site class C (soft rock or very dense soil) can be distinguished by their geologic age and rock type. Site class B may include igneous rocks, metamorphic rocks, limestone, and hard volcanic deposits. Sandstones, shales, conglomerates, and slates of Miocene age or older are all classified as site class B. Pliocene and Pleistocene sandstones, shale/mudstones, and conglomerates are considered to be soft rocks, and thus classified as site class C. Volcanic breccias and pyroclastic rocks of a similar age can also be grouped in site class C. Both the lateritic highlands and the lateritic terraces are classified as site class C. Late Quaternary deposits, other than lateritic terraces, such as loose sand, silt, clay, and gravel deposits, are considered engineering soils and classified as site class D (stiff soil) or site class E (soft soil). Soils in class D are fluvial terraces, stiff clays, and sandy gravel deposits, while Holocene alluvium floodplains or recent fills usually form soils in site class E. For differentiating the stiff soil and soft soil sites, standard penetration test (SPT) values (generally denoted as SPT-N) from borehole data can also be used. The average SPT-N value has to be calculated for the top 30 m of soil, and soils with average SPT-N values greater than 15 are defined as class D; others are defined as class E.

6.3.2 Geotechnical data

Seed and Idriss (1982) developed a site classification system based on geotechnical data by incorporating sediment depth. They classified the soil into four classes: (1) rock sites, (2) stiff soil sites (< 60 m deep), (3) deep cohesionless soil sites (> 75 m deep), and (4) sites underlain by soft to medium stiff clays.

The conventional methods used for geotechnical site classifications are based on the SPT (discussed above) and the cone penetration test (CPT). Of these, the oldest and most widely used method of site classification is based on SPT values. Many correlations are available for evaluating different soil parameters based on the SPT values. The site classification schemes available based on SPT values are given in Table 6.2.

The CPT test also is widely used for evaluation of site classification. Soil classification schemes based on CPT values are given in Table 6.3.

Table 6.2 Soil classification based on SPT-N values

Soil type	N values
Very loose	0–4
Loose	4–10
Medium dense	10–30
Dense	30 –50
Very dense	>50

Source: Budhu (2008).

Table 6.3 Soil classification based on CPT values

Soil type	Cone resistance (q_c) (MPa)
Soft clay and mud	<1
Moderately compact clay	1–5
Silt and loose sand	≤5
Compact to stiff clay and compact silt	>5
Moderately compact sand and gravel	5–12
Compact to very compact gravel	>12

Source: Lunne et al. (1997).

6.3.3   Geophysical data

In recent years, site classification based on the average shear wave velocity values in the top 30 m (V_s^{30}) have been followed by codes such as Eurocode-8 (2003), NEHRP (BSSC, 2003), the International Building Code (IBC, 2009), and others. The amplification of shear waves depends on the density and shear wave velocity of the overlying soil layer. Amplification depends heavily on the shear wave velocity near the earth's surface, as variation in density of the soil bed is relatively less. Two methods are used to denote the near-surface shear wave velocity (V_s): (1) the depth corresponding to a one-quarter wavelength of the period of interest and (2) the average shear wave velocity in the top 30 m. The main disadvantage with the quarter-wavelength V_s is that the depths associated with this will be very deep. Hence, the classification based on V_s^{30} is being used more often. It is calculated using the equation:

$$V_s^{30} = \frac{30}{\sum_{i=1}^{N}\left(\dfrac{d_i}{v_i}\right)} \tag{6.1}$$

where d_i is the thickness of the ith soil layer in meters, v_i is the shear wave velocity for the ith layer, and N is the number of layers in the top 30 m of soil strata that will be considered in evaluating V_s^{30} values.

6.3.4 Eurocode-8 and NEHRP

A site classification scheme based on V_s^{30} values was proposed by Burkhardt (1994), and a similar scheme also was adopted by NEHRP. The NEHRP (BSSC, 2003) site classification scheme is given in Table 6.4. Eurocode-8 (2003) has also classified sites based on V_s^{30}, SPT, and CPT values. The Eurocode-8 classification is provided in Table 6.5. Even though both schemes use similar methods to identify site classes, the range of V_s^{30} values specified for each site class differ.

Table 6.4 Site classification as per the NEHRP scheme

NEHRP site class	Description	V_s^{30} (ms^{-1})
A	Hard rock	>1500
B	Firm and hard rock	760–1500
C	Dense soil, soft rock	360–760
D	Stiff soil	180–360
E	Soft clays	<180
F	Special study soils (e.g., liquefiable soils, sensitive clays, organic soils, soft clays > 36 m thick)	

Source: BSSC (2003).

Table 6.5 Site classification adopted by Eurocode-8 (2003)

Groundtype	Description of stratigraphic profile	Parameters		
		V_s^{30} (ms^{-1})	SPT	C_u (KPa)
A	Rock or other rock-like geological formation, including utmost 5 m of weaker material at the surface.	>800		
B	Deposits of very dense sand, gravel, or very stiff clay, at least several tens of meters in thickness, characterized by a gradual increase of mechanical properties with depth.	360–800	>50	>250
C	Deep deposits of dense or medium dense sand, gravel or stiff clay with a thickness from several tens to many hundreds of meters.	180–360	15–50	70–250
D	Deposits of loose-to-medium cohesionless soil (with or without some soft cohesive layers), or of predominantly soft-to-firm cohesive soil.	<180	<15	<70
E	A soil profile consisting of a surface alluvium layer with V_s^{30} values of type C or D and thickness varying between about 5 m and 20 m, underlain by a stiffer material with V_s^{30} > 800 ms^{-1}.			
S1	Deposits consisting, or containing a layer at least 10 m thick, of soft clays/silts with a high plasticity index (PI > 40) and high water content	<100 (indicative)		10–20
S2	Deposits of liquefiable soils, of sensitive clays, or any other soil profile not included in types A–E or SI			

In many locations, the rock depth will be shallow (less than 30 m), and hence the evaluation of V_s^{30} will not be possible. In those cases, extrapolation of available V_s values has to be done to evaluate the V_s^{30} values. The method proposed by Boore (2004) can be used for this purpose. He has suggested different models to extrapolate the shear wave velocities, for depths less than 30 m, to get the V_s^{30} value. The first method is extrapolation based on constant velocity. In this model, it is assumed that the shear wave velocity remains consistent from the deepest velocity measurement to 30 m:

$$V_S^{30} = \frac{30}{tt(d) + (30-d)/V_{eff}} \tag{6.2}$$

where $tt(d)$ is the travel time to depth d and $V_{eff} = V_s(d), V_s(d)$ is the timed average velocity to a depth of d.

Even though this method is simple, it has been found to underestimate the V_s^{30} values, because in most of the soil the shear wave velocity is found to increase with depth. Another relation proposed by Boore (2004), based on a power law relation, states that V_s^{30} can be estimated as:

$$\log V_s^{30} = a + b \log \bar{V}_s(d) \tag{6.3}$$

where $\bar{V}_s(d)$ is the velocity at a depth of d m ($10 < d < 30$). The values of the regression coefficients a and b can be obtained from Boore (2004). The extrapolation of V_s values can also be determined based on the velocity statistics (Boore, 2004):

$$P(\xi > V_{eff}/V_s(d)) = a(V_{eff}/V_s(d))^b \tag{6.4}$$

where $P(\xi > V_{eff}/V_s(d))$ is the probability of exceedance of $V_{eff}/V_s(d)$. More details of this analysis can be had from Boore (2004).

A modified site classification system based on geotechnical data was proposed by Rodriguez-Marek et al. (2001). In this system, the stiffness of the soil was also taken into account for the site classification. This system is presented in Table 6.6. The main advantage of

Table 6.6 Classification based on geotechnical features

Site	Description	Comments
A	Hard rock	Crystalline bedrock; $V_s^{30} \geq 1500$ ms^{-1}
B	Competent bedrock	$V_s^{30} > 600$ ms^{-1} or < 6 m of soil. Most unweathered California rock cases
C1	Weathered rock	$V_s^{30} \sim 300$ ms^{-1} increasing to > 600 ms^{-1}, weathering zone > 6 m and < 30 m
C2	Shallow stiff soil	Soil depth > 6 m and < 30 m
C3	Intermediate depth stiff soil	Soil depth > 30 m and < 60 m
D1	Deep stiff Holocene soil	Soil depth > 60 m and < 200 m
D2	Deep stiff Pleistocene soil	Soil depth > 60 m and < 200 m
D3	Very deep stiff soil	Soil depth > 200 m
E1	Medium thickness soft clay	Thickness of soft clay layer $3 - 12$ m
E2	Deep soft clay	Thickness of soft clay layer > 12 m
F	Potentially liquefiable sand	Holocene loose sand with high water table, $Z_w \leq 6$ m

Source: Rodriguez-Marek et al. (2001).

this system is that it correlates the V_s^{30} values with the geotechnical and surface geological features.

6.4 DEVELOPMENT OF THE V_s^{30} MAP FROM THE TOPOGRAPHIC SLOPE

Maps of seismic site conditions on regional scales are not readily available as they require extensive geological and geotechnical investigations. Seismic site classification maps are available for only a few citieswhere detailed site investigations have been carried out. Topographic elevation data, in contrast, are available at uniform sampling for the globe. The slope of topography, or gradient, can be indicative of V_s^{30} because rocky materials are more likely to maintain a steep slope, whereas deep basin sediments are deposited in regions with low gradients. The similarity between the topography of California and the surficial site-condition map derived from geology was remarkable (Wills et al., 2000). Recent studies have confirmed good correlations between V_s^{30} and slope in Japan (e.g., Matsuoka et al., 2005) and elevation with V_s^{30} in Taiwan (Chiou and Youngs, 2006). The inspiration for correlating topography and site conditions comes from a practical need to characterize the site in terms of amplification as part of predicting the surface level ground motion. In the present study, an attempt was made to do the site characterization of all of India based on shear wave velocity values estimated from the topographic gradient. It includes two steps: preparation of a slope map and generation of a V_s^{30} map from the slope map.

6.4.1 Preparation of the slope map

The slope map was prepared using satellite images downloaded from various sources: the Cartosat Digital Elevation Model (DEM) from Bhuvan, Shuttle Radar Topography Mission (SRTM) data from USGS, and void-filled data in mountainous terrain from view finder panaromas. Details of the downloaded data are as follows:

1. Universal Transverse Mercator (UTM) grid coverage: 28 grids
2. Total number of data: 390 cells
3. SRTM data: size of cell, 1° × 1°; resolution, 3″
4. Cartosat DEM: size of cell, 1° × 1°; resolution, 1″
5. Mosaic from view finder panaromas: size of cell, 1° × 1° degree; resolution, 3″

The data were resampled and mosaicked to obtain a 6′ resolution map containing the elevation values for the entire region of interest. Negative and spurious data were deleted. The voids (null) were removed and replaced with interpolated values using the spline fitting method to accommodate for surface undulation in mountainous terrain. The slope map of the entire area of interest was generated from the above data using ArcGIS software. The entire country was divided into grids of size 0.1° × 0.1°, the slope value at each grid point was estimated, and the spatial variation of slope values across the country was plotted (Fig. 6.2). Details on the processing of data for generating the slope map are given in Appendix D.

6.4.2 Generation of the V_s^{30} map

Wald and Allen (2007) described a technique to derive site-condition maps directly from topographic data. They used global 30 arc second topographic data and V_s^{30} dimensions collected from various studies in the United States, as well as in Taiwan, Italy, and Australia.

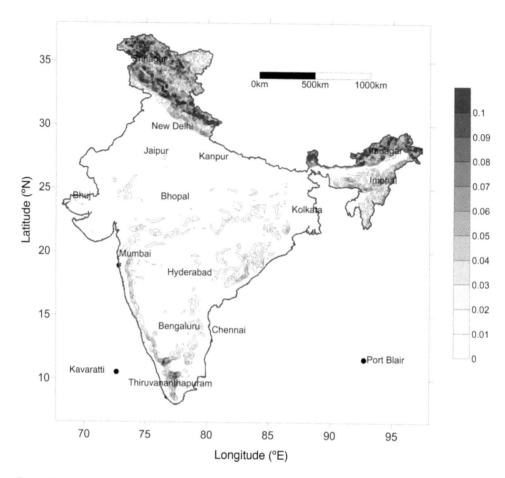

Figure 6.2 Contour map showing the spatial variation of slope value (m m^{-1})

They correlated V_s^{30} values with the topographic slope, developing correlations for obtaining V_s^{30}: one for active tectonic regions where topographic relief is high and one for stable shield regions where the topography is quieter.

By taking the topography gradient and choosing slope ranges that maximized the correlation with shallow shear-velocity observations, Wald and Allen (2007) recovered many of the spatially varying features of the site-condition maps developed in California. They found that maps derived from the slope of the topography correlated well with regional site-condition maps. The correlations proposed by Wald and Allen (2007) are given in Table 6.7. The V_s^{30} contour map developed for India by Sitharam et al. (2015) using these correlations is shown in Figure 6.3 with a comparison of site classification maps developed by various researchers for a number of cities in India using geophysical/geotechnical investigations is presented in Table 6.8.

Table 6.7 Slope ranges for NEHRP V_s^{30} categories

Class	V_s^{30} range (ms^{-1})	Slope range (m m^{-1})	
		Active tectonic	*Stable continent*
E	<180	<1.0 E-4	<2.0 E-5
D	180–240	1.0E-4–2.2E-3	2.0E-5–2.0E-3
	240–300	2.2E-3–6.3E-3	2.0E-3–4.0E-3
	300–360	6.3E-3–0.018	4.0E-3–7.2E-3
C	360–490	0.018–0.050	7.2E-3–0.013
	490–620	0.050–0.10	0.013–0.018
	620–760	0.10–0.138	0.018–0.025
B	>760	>0.138	>0.025

Source: Wald and Allen (2007).

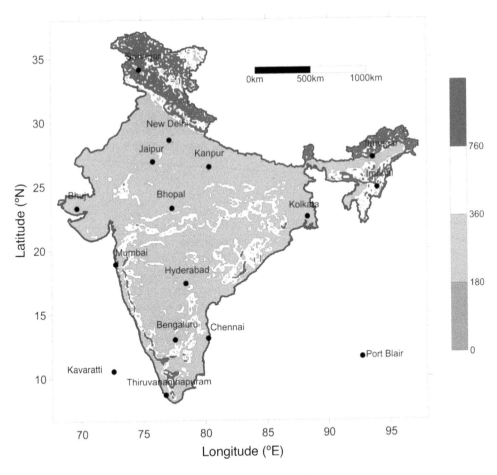

Figure 6.3 V_s^{30} (km s^{-1}) contour map derived from the slope values

Table 6.8 Comparison of site classification of various cities

Location	Longitude (°)	Latitude (°)	Site class	
			Sitharam et al. (2015)	Previous studies
Bengaluru	77.6	13	Site class D	Site class D; Sitharam and Anbazhagan (2008)
Chennai	80.25	13.07	Site class D	Site class D; Boominathan et al. (2008)
New Delhi	77.20	28.58	Site class D	Site class D; Rao and Ramona (2008)
Impala	93.95	24.80	Site class D	Site class D; (Pallav et al.,2015)

6.5 ESTIMATION OF SURFACE-LEVEL PGA VALUES

Two approaches in estimating the surface-level PGA values are discussed here. As a first-level approach, four contour maps of surface-level PGA, corresponding to four different NEHRP site classes, were prepared for the entire country. For any location in the country, the site class can be identified through a site-specific investigation, and the surface-level PGA value can be found from the map corresponding to that site class. Second, a single contour map of surface-level PGA was prepared after characterizing the entire country into different NEHRP site classes based on V_s^{30} values estimated from the slope map. Both methods are discussed below.

The evaluation of PGA values for different site classes was done using the amplification factor equations suggested by Raghu Kanth and Iyengar (2007). The amplification factors for different NEHRP site classes can be evaluated using the following equation (Raghu Kanth and Iyengar, 2007):

$$F_s = a_1 \, y_{br} + a_2 + \ln is$$

(6.5)

Where a_1 and a_2 are regression coefficients, y_r is the spectral acceleration at rock level, and is is the error term. The values of the regression coefficients a_1 and a_2 will vary for different site classes and for different time periods. These values were derived based on the statistical simulation of ground motions (Raghu Kanth and Iyengar, 2007), and they take into account the nonlinear site response of soils. For NEHRP site classes A to D, 10 random samples of soil profiles were considered in evaluating the amplification factors. The values of a_1, a_2, and δ for different site classes that were used to evaluate the PGA values are given in Table 6.9. The amplification factor for soft and medium-dense soil varies with the rock-level

Table 6.9 Amplification factors for site classes

Site class	a_1	a_2	$\ln \delta_s$
A	0	0.36	0.03
B	0	0.49	0.08
C	−0.89	0.66	0.23
D	−2.61	0.80	0.36

Source: Raghu Kanth and Iyengar (2007).

PGA values. The value of the damping ratio, ξ, and the modulus reduction, G_{red}, will vary with shear strain (Idriss, 1990). The method adopted for evaluation of amplification factor (F_s) values considers this effect, and the value of F_s varies with the rock-level PGA values. The value of spectral acceleration for different site classes can be obtained from:

$$Yes = y_{rs}.$$ (6.6)

F_s is the amplification factor and is the spectral acceleration at the ground surface for a given site class. The value of F_s varies with PGA values, and it accounts for the nonlinear behaviour of soil in amplifying the seismic signals.

In the present study, the surface-level PGA values were evaluated for India based on the rock-level PGA values obtained from both deterministic and probabilistic analysis. The study area covers more than 3 million km², and it is challenging to classify the region into different site classes based on the geotechnical or geophysical data. Hence, first the PGA values were evaluated for all four NEHRP site classes (A to D). Based on site investigation one can identify the site class, and for the appropriate site class the PGA or Sa values can be obtained from the respective map. The PGA values were evaluated for four NEHRP site classes (A to D) based on the amplification factors suggested by Raghu Kanth and Iyengar (2007).

The spatial variation of surface-level PGA values for different site classes obtained from DSHA is given in Figure 6.4. The same obtained using PSHA for a return period of 475 years

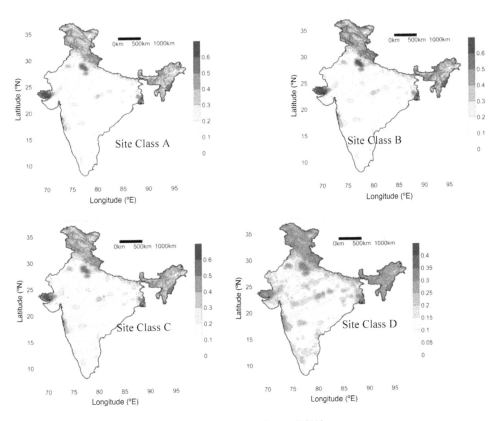

Figure 6.4 Surface-level PGA maps for different site classes: DSHA

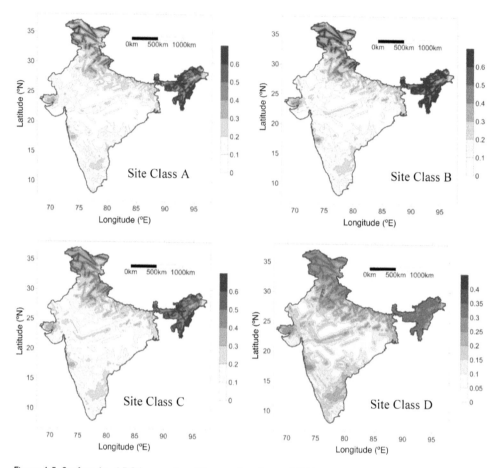

Figure 6.5 Surface-level PGA maps for different site classes: PSHA

is shown in Figure 6.5. These results clearly show the effect of amplification due to the overlying soil column. The rock-level PGA values indicate that for the majority of the study area the PGA value at bedrock was less than 0.4 g. However, when the site effects were also considered, especially for site class C (medium-dense soil), the PGA values for the majority of the study area increased considerably. The amplification for higher PGA values was more for site classes A and B and the amplification for lower values of PGA were more in site classes C and D. The spectral acceleration values were also evaluated for different site classes. For all the site classes the maximum PGA values were obtained in regions belonging to north and northeast India. A geotechnical site investigation will indicate the site class (based on SPT, CPT, or V_s^{30}) at the desired location. Depending on the site class to which the place belongs, the PGA at ground surface can be obtained from the respective values. However, for many locations the SPT, CPT, or V_s^{30} values may not be available; for those areas the site classification can be done based on the local geology and then the corresponding NEHRP site class can be determined. Thus, it provides a straightforward and comprehensive method to obtain the PGA value at ground surface for a vast country like India.

Given that it is difficult to do geotechnical/geophysical site investigation for a vast country like India, we have used the V_s^{30} map generated from the slope values, as discussed in section 6.4. Using appropriate amplification factors corresponding to shear wave velocity values, we generated a single map for all of India showing the spatial variation of surface-level PGA. This map can be used to estimate surface-level PGA at any location in the country. The maps were developed from the PGA values obtained from both deterministic and probabilistic approaches and are presented in Figures 6.6 and 6.7, respectively. It can be seen that PGA values are significantly amplified in regions that belong to site class C. Most of India belongs to site class D, and regions of low PGA values at bedrock level show significant amplification at surface level. Regions of high PGA values at bedrock level (e.g., Indo-Myanmar subduction region) show deamplification of the PGA at the surface level. The present surface-level PGA values match well with those obtained for various cities whose microzonation studies have been carried out earlier (Table 6.10). The PGA obtained at ground level using this approach is a simple method and a good first approximation that may be used for preliminary design.

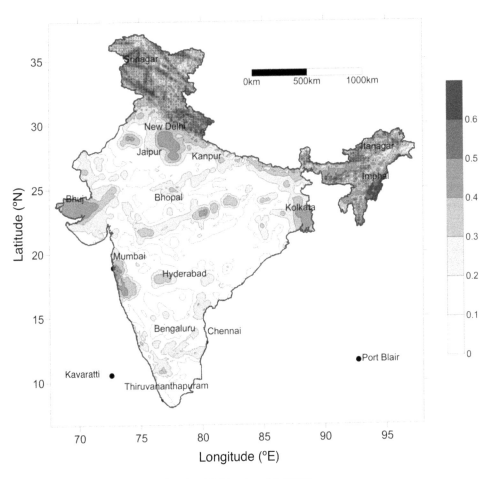

Figure 6.6 Spatial variation of surface-level PGA values (g): DSHA

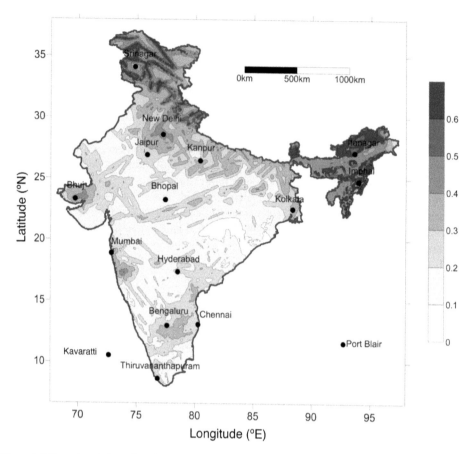

Figure 6.7 Spatial variation of surface-level PGA values (g) for a return period of 475 years: PSHA
Source: Sitharam et al. (2013).

Table 6.10 Comparison of surface-level PGA values with earlier studies

Major cities	Location		PGA (g)		
	Longitude (°E)	Latitude (°N)	Present DSHA	Present PSHA (475-year return period)	Previous PSHA studies
Gangtok	88.4	27.2	0.6	0.702	0.6 (Nath et al., 2008)
Delhi	77.21	28.61	0.45	0.42	>0.24 (Iyengar and Ghosh,2004)
Bangalore	77.6	13.0	0.237	0.3	0.3–0.45 (Anbazhagan et al., 2009)
Guwahati	91.6	26.2	0.45	0.4	0.45 (Nath et al., 2008)
Chennai	80.2	13.1	0.16	0.28	0.141 (Boominathan et al., 2008)
Imphal	93.95	24.8	0.53	0.63	0.5 (Pallav et al., 2015)

6.6 EVALUATION OF LIQUEFACTION POTENTIAL

The important steps involved in the evaluation of liquefaction potential are as follows:

- Evaluation of peak horizontal acceleration (PGA) at bedrock level.
- Evaluation of surface-level PGA, considering site effects.
- Evaluation of liquefaction potential based on the PGA values and the soil properties.

The conventional liquefaction evaluation methods use single ground acceleration and earthquake magnitude values. The evaluation of seismic hazard using PSHA shows that particular ground acceleration was not contributed by a single earthquake magnitude; instead, it was contributed by different magnitudes with varying probability of occurrence. A new probabilistic performance-based approach based on SPT values, suggested by Kramer and Mayfield (2007), utilizes the entire ground acceleration range in evaluating the liquefaction potential. They evaluated the liquefaction return period based on SPT values in India by considering uncertainties in earthquake loading. In the present study, the liquefaction potential is evaluated using a probabilistic performance-based approach by using a logic tree method.

The liquefaction potential of the Indian landmass was evaluated based on SPT values. The evaluation of liquefaction potential involves two stages: (1) evaluation of earthquake loading and (2) evaluation of soil strength against earthquake loading. The earthquake loading on soil is expressed using the cyclic stress ratio (CSR), and the soil strength (capacity of soil) to resist liquefaction is expressed using the cyclic resistance ratio (CRR). One of the first methods to evaluate earthquake loading, CSR was suggested by Seed and Idriss (1971) in the "simplified method":

$$CSR = 0.65 \frac{a_{max}}{g} \frac{\sigma_{vo}}{\sigma'_{vo}} \frac{r_d}{MSF} \tag{6.7}$$

where a_{max} is the PGA (at surface level); σ_{vo} and σ'_{vo} are the total and effective overburden pressure, respectively; r_d is the depth reduction factor used to account for the flexibility of the soil; and MSF is the magnitude scaling factor. This relationship was developed for an earthquake of magnitude $M_W = 7.5$, and if the magnitude of an earthquake is different from this, it is taken care of by the MSF. The most widely used methods to evaluate the CRR value based on SPT values were proposed by Seed et al. (1985). This relationship is between the corrected SPT-N values (corrected for overburden stress, instrument errors, and other factors that affect the SPT testing) and the intensity of cyclic loading that is expressed in the form of a uniform CSR. This relation was further modified by Youd and Idriss (2001). Even though these relations are widely used, they do not have a formal probabilistic approach; that is, the relationship does not provide the probability or uncertainty of occurrence of liquefaction.

A number of researchers have developed correlations based on probabilistic methods to evaluate the liquefaction potential, and one of the first attempts was made by Liao et al. (1988). Recent and comprehensive work in this area was done by Cetin et al. (2004).

The incorporation of earthquake loading into evaluation of liquefaction potential requires the quantification of the uncertainties in earthquake loading. All the available methods, either probabilistic or deterministic, use a single ground acceleration and earthquake magnitude.

The results obtained from PSHA show that several magnitudes contribute to ground acceleration, and the percentage of contribution varies. It would not be fair to conclude that a particular ground acceleration was produced by a particular magnitude; instead, different magnitudes contribute to it. The conventional liquefaction analysis methods fail to consider this aspect of earthquake loading. Moreover, the annual frequency of occurrence of lower acceleration values will be more, and that of higher acceleration values will be less. Conventional liquefaction analysis also fails to account for such variations in the frequency of occurrence of ground motions. Hence, to account for these uncertainties in a better way, a probabilistic performance-based liquefaction potential evaluation method was suggested by Kramer and Mayfield (2007). According to this method, the annual rate of exceedance of engineering design parameter can be obtained by:

$$\lambda_{EDP^*} = \sum_{i=1}^{N_{IM}} P\left[EDP > EDP^* \tfrac{1}{2} IM = im_i\right] \Delta\lambda_{im_i} \tag{6.8}$$

where EDP is the engineering design parameter say factor of safety; EDP^* is a selected value of EDP; IM is the intensity measure that is used to characterize earthquake loading say peak ground acceleration; im_i is the discretized value of IM; λ_{EDP^*} is the mean annual rate of exceedance of EDP^*; and $\Delta\lambda_{im_i}$ is the incremental mean annual rate of exceedance of the discretized value of the intensity measure im.

Equation 6.9 can be obtained by considering EDP as the factor of safety and the intensity measure of ground motion as the earthquake ground acceleration:

$$\Lambda_{FS_L^*} = \sum_{i=1}^{N_a} \sum_{j=1}^{N_M} P[FS_L < FS_L^* \mid a_i, m_j] \Delta\lambda_{a_i}, m_j \tag{6.9}$$

where $\Lambda_{FS_L^*}$ is the annual rate at which the safety factor will be less than $FS^*{}_L$; N_M is the number of magnitude increments; N_a is the number of peak acceleration increments; and $\Delta\lambda_{a_i}, m_j$ is the incremental annual frequency of exceedance for acceleration a_i and magnitude m_j (this value can be obtained from the deaggregated seismic hazard curve with respect to magnitude). The conditional probability in the previous equation can be written as (Kramer and Mayfield, 2007):

$$P[FS_L < FS_L^* \mid a_i, m_j] = \Phi\left[-\frac{\begin{array}{l}(N_1)_{60}(1 + \theta_1 FC) - \theta_2 \ln(CSR_{eq,i} FS_L^*) - \\ \theta_3 \ln(m_j) - \theta_4 (\ln(\sigma_{v0}^t / P_a) + \theta_5 FC + \theta_6\end{array}}{\sigma_\varepsilon}\right] \tag{6.10}$$

$$CSR_{eq,i} = 0.65 \frac{a_i}{g} \frac{\sigma_{vo}}{\sigma'_{vo}} r_d \tag{6.11}$$

where $CSR_{eq,i}$ is the CSR value calculated without using the MSF for an acceleration a_i, and this will be calculated for all the acceleration levels. The most widely used technique to estimate the stress reduction factor (r_d) was suggested by Seed and Idriss (1971). Further, Cetin and Seed (2004) evolved a method to evaluate the stress reduction factor as a function of depth, earthquake magnitude, ground acceleration, and the average shear wave velocity of the top 12 m of the soil column. For a depth less than 20 m, the value of r_d is given by:

$$r_d(d, Mw, a_{max}, V_{s,12}^*) = \frac{1 + \dfrac{-23.013 - 2.949a_{max} + 0.999M_w + 0.0525V_{s,12}^*}{16.258 + 0.201e^{0.341(-d+0.0785V_{s,12}^*+7.586)}}}{1 + \dfrac{-23.013 - 2.949a_{max} + 0.999M_w + 0.0525V_{s,12}^*}{16.258 + 0.201e^{0.341(0.0785V_{s,12}^*+7.586)}}} \pm \sigma_{\varepsilon_{rd}} \quad (6.12)$$

where a_{max} and M_W are the maximum acceleration (in g) and corresponding earthquake moment magnitude values; $V_{s,12}^*$ is the average shear wave velocity in ms^{-1} for the top 12 m of the soil layer; and $\sigma_{\varepsilon_{rd}}$ is the standard deviation of model error.

For the evaluation of liquefaction potential, the surface-level PGA values presented in the above section were used. Only the regions that fall under site class D were considered for liquefaction study, as the other site classes (A, B, or C) denote very stiff soil and rock, which will not liquefy. In order to consider the worst-case scenario for liquefaction, the water table was assumed to be at ground level. The unit weight of the soil was taken as 18 kN m^{-3}, and the corrected SPT $(N_1)_{60,cs}$ values required to prevent liquefaction were evaluated at a depth of 3 m. The liquefaction hazard curves showing the $(N_1)_{60,cs}$ values needed to avert liquefaction against the annual frequency of exceedance (inverse of return period) were developed for each of the grid points, using the methods explained in the previous sections. These curves showed the variation of $(N_1)_{60,cs}$ values required to prevent liquefaction against the annual frequency of exceedance. The corrected SPT values at that location can be obtained from a field test, and if the values obtained after applying all the required corrections are higher than those values obtained from the hazard curve, then it means that the location is safe against liquefaction for that particular return period. In addition to this, the return period corresponding to the corrected SPT value obtained from the field will give the return period of seismic soil liquefaction at that location. Hence, this method provides a very comprehensive and efficient way of determining the liquefaction return period of a very vast area, where the geotechnical properties are not available. The corrected SPT values required to prevent liquefaction of the selected locations with different probabilities of exceedance in 50 years are shown in Table 6.11. These values indicate the $(N_1)_{60,cs}$ values required to prevent liquefaction for the specified return period. If the corrected SPT values at the site

Table 6.11 SPT values required to prevent liquefaction at different cities for different return periods

Major cities	Location		Surface-level PGA value (g)	Required SPT value	
	Longitude (°E)	Latitude (°N)		For return period of 475 years	For return period of 2475 years
Mumbai	72.82	18.90	0.20	8	17
Delhi	77.20	28.58	0.42	23	25
Bangalore	77.59	12.98	0.30	12	21
Kolkata	88.33	22.53	0.29	13	22
Chennai	80.25	13.07	0.28	9	15

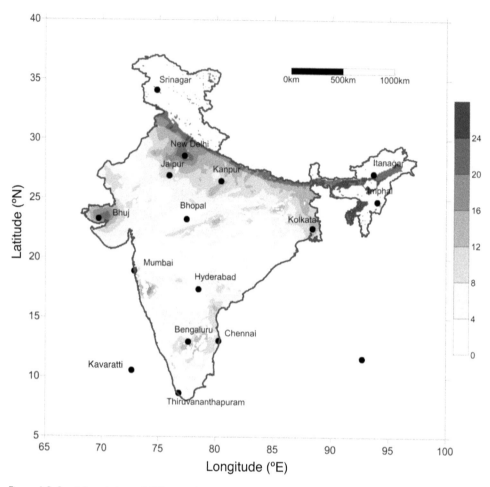

Figure 6.8 Spatial variation of $(N_1)_{60,cs}$ values required to prevent liquefaction for a return period of 475 years

(obtained from a site investigation) are higher than the values given in the table, then these locations are safe against liquefaction for that given return period; otherwise, these sites are vulnerable to liquefaction hazard. The liquefaction hazard curves were generated for the entire study area. From these curves, the corrected $(N_1)_{60,cs}$ values required to prevent lique-faction for 10% and 2% probability of exceedance in 50 years were evaluated for the entire study area (this corresponds to a return period of 475 and 2475 years, respectively).

The contour curves showing the spatial variation of $(N_1)_{60,cs}$ values required to prevent liquefaction for return periods of 475 and 2475 years are shown in Figures 6.8 and 6.9. The white (blank) regions shown in these figures do not belong to site class D, and hence the probability of liquefaction is zero.

The patterns of spatial variation of $(N_1)_{60,cs}$ values are similar for both return periods. The $(N_1)_{60,cs}$ values required to prevent liquefaction for a return period of 2475 years are higher due to the increase in return period and the subsequent increase in the PGA values. The

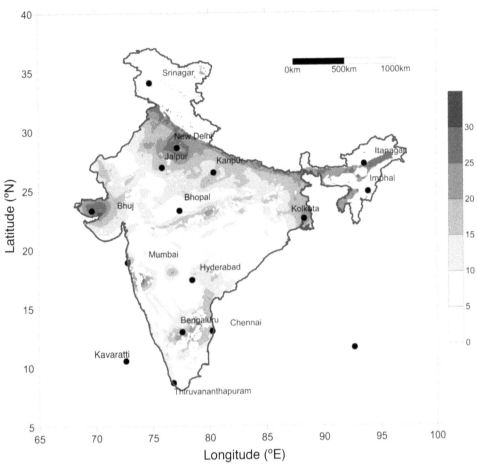

Figure 6.9 Spatial variation of $(N_1)_{60,cs}$ values required to prevent liquefaction for a return period of 2475 years

actual $(N_1)_{60,cs}$ values at a particular site can be obtained from geotechnical investigation. If the $(N_1)_{60,cs}$ value obtained is greater than the values specified in Figures 6.8 and 6.9, then the location is safe against liquefaction for that return period. For any region in the study area, the factor of safety against liquefaction for a given return period can be obtained by dividing the values presented in this study with the actual $(N_1)_{60,cs}$ values. In a similar way, liquefaction susceptibility maps also can be prepared for another return period.

6.7 GENERAL REMARKS

Site response studies and the estimation of surface-level PGA values is of very high importance in engineering design. This chapter discussed evaluation of surface-level PGA values for Indian subcontinent for four NEHRP site classes by considering local site effects. In the absence of V_s^{30} values, the site classes can be identified based on local geological conditions. This chapter also presented V_s^{30} characterizations for India based on the topographic gradient. Thus, the PGA value was brought to the surface level using appropriate amplification factors.

Surface-level PGA values were evaluated for four site classes by considering local site effects. If the site class at any location in the study area is known, then the ground-level PGA values can be obtained from the respective map. Thus, this method provides a simplified methodology for evaluating surface-level PGA values. PGA values obtained based on DSHA can be taken as the upper-bound PGA values for the respective site classes in India. The limitation of this work is that amplification effects due to surface topography and basin geometry were not considered. Also, the final surface-level PGA map was generated based on a V_s^{30} map created from slope values using correlations, not on a geotechnical investigation. The spatial variation of SPT values required to prevent liquefaction plotted using a probabilistic methodology will act as a guideline to understand the liquefaction susceptibility of any location within the study region.

References

Abrahamson, N.A. & Silva, W.J. (1997) Empirical response spectral attenuation relations for shallow crustal earthquakes. *Seismological Research Letters*, 68, 94–127.

Abramowitz, M. & Stegun, I.A. (1970) *Handbook of Mathematical Functions*. 9th edn. Dover Publishers, New York.

Aki, K. (1965) Maximum likelihood estimate of b in the formula log N = a-bm and its confidence limits. *Bulletin of Earthquake Research Institute, University of Tokyo*, 43, 237–239.

Aki, K. (1966) Generation and propagation of G waves from the Niigata earthquake of June 16, 1964: 2. Estimation of earthquake movement, released energy, and stress-strain drop from G wave spectrum. *Bulletin of Earthquake Research Institute*, 44, 23–88.

Aki, K. (1969) Analysis of seismic coda of local earthquakes as scattered waves. *Journal of Geophysical Research*, 74, 615–631.

Akkar, S. & Bommer, J.J. (2010) Empirical equations for the prediction of PGA, PGV, and spectral accelerations in Europe, the Mediterranean region, and the Middle East. *Seismological Research Letters*, 81, 195–206.

Algermissen, S.T. & Perkins, D.M. (1976) A probabilistic estimate of maximum acceleration in rock in the contiguous United States. *U.S. Geological Survey: Open-File Report*, 76–416, 45pp.

Anbazhagan, P., Bajaj, K., Dutta, N., Moustafa, S.S. & Al-Arifi, N.S. (2017) Region-specific deterministic and probabilistic seismic hazard analysis of Kanpur city. *Journal of Earth System Science*, 126(1), 12.

Anbazhagan, P., Bajaj, K. & Patel, S. (2015) Seismic hazard maps and spectrum for Patna considering region-specific seismotectonic parameters. *Natural Hazards*, 78(2), 1163–1195.

Anbazhagan, P., Smitha, C.V. & Kumar, A. (2014) Representative seismic hazard map of Coimbatore, India. *Engineering Geology*, 171, 81–95.

Anbazhagan, P., Vinod, J.S. & Sitharam, T.G. (2009) Probabilistic seismic hazard analysis for Bangalore. *Journal of Natural Hazards*, 48, 145–166.

Anderson, J.G., Wesnousky, S.G. & Stirling, M.W. (1996) Earthquake size as a function of slip rate. *Bulletin of the Seismological Society of America*, 86, 683–690.

Atkinson, G.M. & Boore, D.M. (2003) Empirical ground-motion relations for subduction-zone earthquakes and their application to Cascadia and other regions. *Bulletin of the Seismological Society of America*, 93, 1703–1729.

Atkinson, G.M. & Boore, D.M. (2006) Earthquake ground-motion prediction equations for Eastern North America. *Bulletin of the Seismological Society of America*, 96(6), 2181–2205.

Basu, K.L. (1964) A note on the Coimbatore earthquake of 8th February 1900. *Indian Journal of Meteorology and Geophysics*, 15(2), 281–286.

Bennington, N., Thurber, C. & Roecker, S. (2008). Three-dimensional seismic attenuation structure around the SAFOD site, Parkfield, California. *Bulletin of the Seismological Society of America*, 98(6), 2934–2947.

Bhatia, S.C., Kumar, M.R. & Gupta, H.K. (1999) A probabilistic seismic hazard map of India and adjoining regions. *Annali di Geofisica*, 42, 1153–1166.

Bilham, R. (2004) Earthquakes in India and the Himalaya: Tectonics, geodesy and history. *Annals of Geophysics*, 47, 839–858.

Bilham, R., Gaur, V.K. & Molnar, P. (2001) Himalayan seismic hazard. *Science*, 293(5534), 1442–1444.

BIS-1893 (2002) *Indian Standard Criteria for Earthquake Resistant Design of Structures, Part 1: General Provisions and Buildings*. Bureau of Indian Standards, New Delhi.

Biswas, S. & Dasgupta, A. (1986) Some observations on the mechanism of earthquakes in the Himalaya and the Burmese arc. *Tectonophysics*, 122, 325–343.

Bodin, P., Malagnini, L. & Akinci, A. (2004) Ground-motion scaling in the Kachchh Basin, India, deduced from aftershocks of the 2001 MW 7.6 Bhuj Earthquake. *Bulletin of the Seismological Society of America*, 94, 1658–1669.

Bommer, J.J., Scherbaum, F., Bungum, H., Cotton, F., Sabetta, F. & Abrahamson, N.A. (2005) On the use of logic trees for ground-motion prediction equations in seismic hazard analysis. *Bulletin of the Seismological Society of America*, 95, 377–389.

Boominathan, A., Dodagoudar, G.R., Suganthi, A. & Uma Maheswari, R. (2008) Seismic hazard assessment of Chennai city considering local site effects. *Journal of Earth System Sciences*, 117(S2), 853–863.

Boore, D.M. (2004) Estimating vs (30) (or NEHRP Site Classes) from shallow velocity models (depths < 30 m). *Bulletin of the Seismological Society of America*, 94(2), 591–597.

Boore, D.M. & Atkinson, G.M. (2008) Ground-motion prediction equations for the average horizontal component of PGA, PGV, and 5%-damped PSA at spectral periods between 0.01 s and 10.0 s. *Earthquake Spectra*, 24, 99–138.

Borcherdt, R.D. (1994) Estimates of site-dependent response spectra for design (methodology and justification). *Earthquake Spectra*, 10(4), 617–653.

Bozzoni, F., Corigliano, M., Lai, C.G., Salazar, W., Scandella, L., Zuccolo, E., Latchman, J., Lynch, L. & Robertson, R. (2011) Probabilistic seismic hazard assessment at the Eastern Caribbean Islands. *Bulletin of the Seismological Society of America*, 101(5), 2499–2521.

Briggs, J.C. (2003) The biogeographic and tectonic history of India. *Journal of Biogeography*, 30, 381–388.

BSSC (2003) NEHRP recommended provisions for seismic regulations for new buildings and other structures (FEMA 450), Part 1: Provisions. *Building Seismic Safety Council for the Federal Emergency Management Agency*. Washington, DC, USA.

Budhu, M. (2008) *Soil Mechanics and Foundations*. 3rd edn. John Wiley and Sons, Inc., New York.

Budnitz, R.J., Apostolakis, G. & Boore, D.M. (1997). Recommendations for probabilistic seismic hazard analysis: guidance on uncertainty and use of experts (No. NUREG/CR-6372-Vol. 1; UCRL-ID-122160). *Nuclear Regulatory Commission*, Washington, DC (United States). Div. of Engineering Technology; Lawrence Livermore National Lab., CA (United States); Electric Power Research Inst., Palo Alto, CA (United States); USDOE, Washington, DC (United States).

Campbell, K.W. & Bozorgnia, Y. (2003) Updated near-source ground motion (attenuation) relations for the horizontal and vertical components of peak ground acceleration and acceleration response spectra. *Bulletin of the Seismological Society of America*, 93, 314–331.

Cassidy, J.F. (2013) Seismic gap. In: Bobrowsky, P.T. (ed.) *Encyclopedia of Natural Hazards*. Encyclopedia of Earth Sciences Series. Springer, Dordrecht.

Castro, R.R., Condori, C., Romero, O., Jacques, C. & Suter, M. (2008). Seismic attenuation in northeastern Sonora, Mexico. *Bulletin of the Seismological Society of America*, 98(2), 722–732.

Cetin, K.O. & Seed, R.B. (2004) Non linear shear mass participation factor (rd) for cyclic shear stress ratio evaluation. *Soil Dynamics and Earthquake Engineering*, 24, 103–113.

Cetin, K.O., Seed, R.B., Kiureghian, D.A., Tokimatsu, K., Harder, L.F., Kayen, R.E. & Moss, R.E.S. (2004) Standard penetration test-based probabilistic and deterministic assessment of seismic soil liquefaction potential. *Journal of Geotechnical and Geoenvironmental Engineering*, 130(12), 1314–1340.

Chandra, U. (1977) Earthquakes of Peninsula India-a seismotectonic study. *Bulletin of the Seismological Society of America*, 67, 1387–1413.

Chandra, U. (1978) Seismicity, earthquake mechanisms along the Himalayan mountain range and vicinity. *Physics of the Earth's Interior*, 16, 109–131.

Chernick, M.R. (1999) *Bootstrap Methods: A Practitioner's Guide*. Wiley Series in Probability and Statistics. Wiley, New York.

Chiou, B.S.J. & Youngs, R.R. (2006) PEER-NGA empirical ground motion model for the average horizontal component of peak acceleration and pseudo-spectral acceleration for spectral periods of 0.01 to 10 seconds. *Interim Report for USGS Review*, 219.

Chiou, B.J. & Youngs, R.R. (2008). An NGA model for the average horizontal component of peak ground motion and response spectra. *Earthquake Spectra*, 24(1), 173–215.

Chun, K.Y. & Henderson, G.A. (2009). Lg attenuation near the North Korean border with China, Part II: Model development from the 2006 nuclear explosion in North Korea. *Bulletin of the Seismological Society of America*, 99(5), 3030–3038.

Cornell, C.A. (1968) Engineering seismic risk analysis. *Bulletin of the Seismological Society of America*, 58, 1583–1606.

Costa, G., Panza, G.F., Suhadolc, P. & Vaccari, F. (1993). Zoning of the Italian territory in terms of expected peak ground acceleration derived from complete synthetic seismograms. *Journal of Applied Geophysics*, 30(1–2), 149–160.

Cramer, C.H. & Kumar, A. (2003) 2001 Bhuj, India, earthquake engineering seismoscope recordings and Eastern North America ground motion attenuation relations. *Bulletin of the Seismological Society of America*, 93, 1390–1394.

Cramér, H. (2016). Mathematical methods of statistics (PMS-9) (Vol. 9). Princeton university press.

Curray, J.R. (2005) Tectonics and history of the Andaman Sea region. *Asian Journal of Earth Sciences*, 25, 187–228.

Das, R., Sharma, M.L. & Wason, H.R. (2016) Probabilistic seismic hazard assessment for Northeast India Region. *Pure and Applied Geophysics*, 173(8), 2653–2670.

Dasgupta, S. & Mukhopadhyay, M. (1993) Seismicity and plate deformation below the Andaman Arc, Northeast Indian Ocean. *Tectonophysics*, 225, 529–542.

Dasgupta, S. & Mukhopadhyay, M. (1997) Aseismicity of the Andaman subduction zone and recent volcanism. *Journal of Geological Society of India*, 49, 513–521.

Dasgupta, S., Mukhopadhyay, B. & Acharyya, A. (2007a) Seismotectonic of Andaman-Nicobar region: Constraints from aftershocks within 24 hours of the Great 26 December 2004 earthquake. *Geological Survey of India Special Publication*, 89, 95–104.

Dasgupta, S., Mukhopadhyay, B. & Bhattacharya, A. (2007b) Seismicity pattern in north Sumatra: Great Nicobar region: In search of precursor for the 26 December 2004 earthquake. *Journal of Earth System Science*, 116(3), June, 215–223.

Dasgupta, S., Mukhopadhyay, M., Bhattacharya, A. & Jana, T.K. (2003) The geometry of the Burmese: Andaman subducting lithosphere. *Journal of Seismology*, 7, 155–174.

Dasgupta, S., Pande, P., Ganguly, D., Iqbal, Z., Sanyal, K., Venkatraman, N.V. et al. (2000) *Seismotectonic Atlas of India and Its Environs*. Geological Survey of India, Calcutta.

Davis, S.D. & Frohlich, C. (1991) Single-link cluster analysis, synthetic earthquake catalogs and aftershock identification. *Geophysical Journal International*, 104, 289–306.

Day, R.W. (2002) *Geotechnical Earthquake Engineering Handbook*. McGraw-Hill, New York.

Desai, S.S. & Choudhury, D. (2014) Spatial variation of probabilistic seismic hazard for Mumbai and surrounding region. *Natural Hazards*, 71(3), 1873–1898.

Deshikachar, S.V. (1974) A review of the tectonic and geological history of eastern India in terms of "plate tectonics" theory. *Journal of Geological Society of India*, 15, 137–149.

Dhar, S., Rai, A.K. & Nayak, P. (2017) Estimation of seismic hazard in Odisha by remote sensing and GIS techniques. *Natural Hazards*, 86(2), 695–709.

Dunbar, P.K., Lockridge, P.A. & Whiteside, L.S. (1992) Catalog of significant earthquakes, 2150 B.C.– 1991 A.D. *NOAA/NGDC Report SE-49*, Boulder, Colorado, 320p.

Eguchi, T., Uyeda, S. & Maki, T. (1979) Seismotectonics and tectonic history of Andaman sea. *Tectonophysics*, 57, 35–51.

Eurocode-8 (2003) *BS-EN 1998–1, Design of Structures for Earthquake Resistance, Part 1: General Rules, Seismic Actions and Rules for Buildings*. European Committee for Standardization, Brussels.

Evans, P. (1964) The tectonic framework of Assam. *Journal of Geological Society of India*, 5, 80–96.

Field, E.H., Jackson, D.D. & Dolan, J.F. (1999) A mutually consistent seismic-hazard source model for Southern California. *Bulletin of the Seismological Society of America*, 89, 559–578.

Ford, S.R., Dreger, D.S., Mayeda, K., Walter, W.R., Malagnini, L. & Phillips, W.S. (2008). Regional attenuation in northern California: A comparison of five 1D Q methods. *Bulletin of the Seismological Society of America*, 98(4), 2033–2046.

Frank, S., Bommer, J.J., Bungum, H., Cotton, F. & Abrahamson, N.A. (2005) Composite ground-motion models and logic trees: Methodology, sensitivities, and uncertainties. *Bulletin of the Seismological Society of America*, 95(5), 1575–1593.

Frankel, A. (1995) Mapping seismic hazard in the Central Eastern United States. *Seismological Research Letters*, 66(4), 8–21.

Ganesha Raj, K. & Nijagunappa, R. (2004) Major lineaments of Karnataka state and their relation to seismicity: Remote sensing based analysis. *Journal of the Geological Society of India*, 63, 430–439.

Gangrade, B.K. & Arora, S.K. (2000) Seismicity of the Indian Peninsular Shield from regional earthquake data. *Pure and Applied Geophysics*, 157, 1683–1705.

Gansser, A. (1964) *Geology of the Himalayas*. Interscience, New York.

Gardner, J.K. & Knopoff, L. (1974) Is the sequence of earthquakes in Southern California with aftershocks removed, Poissonian? *Bulletin of the Seismological Society of America*, 64(5), 1363–1367.

Ghasemi, H., Zare, M., Fukushima, Y. & Koketsu, K. (2009). An empirical spectral ground-motion model for Iran. *Journal of Seismology*, 13(4), 499–515.

Giardini, D. (1984) Systematic analysis of deep seismicity: 200 centroid-moment tensor solutions for earthquakes between 1977 and 1980. *Geophysical Journal Royal Astronomical Society*, 77, 883–914.

Giardini, D., Grünthal, G., Shedlock, K.M. & Zhang, P. (1999) The global seismic hazard assessment program (GSHAP). *Anna Geofisica*, 42, 1225–1228.

Graizer, V. (2016) Ground-motion prediction equations for central and eastern North America. *Bulletin of the Seismological Society of America*, 106(4), 1600–1612.

Gregor, N.J., Silva, W.J., Wong, I.G. & Youngs, R. (2002) Ground motion attenuation relationships for Cascadia subduction zone mega-thrust earthquakes based on a stochastic finite-fault modeling. *Bulletin of the Seismological Society of America*, 92, 1923–1932.

Guha, S.K. & Basu, P.C. (1993) Catalogue of earthquakes ($M \geq 3.0$) in Peninsular India. *Atomic Energy Regulatory Board, Tech*, Document No. TD/CSE-1, 1–70pp.

Gupta, I. D. (2002). The state of the art in seismic hazard analysis. *ISET Journal of Earthquake Technology*, 39(4), 311–346.

Gupta, I.D. (2006) Delineation of probable seismic sources in India and neighborhood by a comprehensive analysis of seismotectonic characteristics of the region. *Soil Dynamics and Earthquake Engineering*, 26, 766–790.

Gupta, I.D. (2010) Response spectral attenuation relations for intraslab earthquakes in Indo-Burmese subduction zone. *Soil Dynamics and Earthquake Engineering*, 30, 368–377.

Gutenberg, B. (1945) Amplitudes of P, PP, and S and magnitude of shallow earthquakes. *Bulletin of the Seismological Society of America*, 35, 57–69.

Gutenberg, B. & Richter, C.F. (1936) Magnitude and energy of earthquakes. *Science*, 83(2147), 183–185.

Gutenberg, B. & Richter, C.F. (1944) Frequency of earthquakes in California. *Bulletin of the Seismological Society of America*, 34, 185–188.

Gutenberg, B. & Richter, C.F. (1956) Magnitude and energy of earthquakes. *Annali De Geofisica*, 9, 1–15.

Hanumantharao, C. & Ramana, G.V. (2008). Dynamic soil properties for microzonation of Delhi, India. *Journal of earth system science*, 117(2), 719–730.

Heaton, T.H., Tajima, T.F. & Mori, A.W. (1986) Estimating ground motions using recorded accelerograms. *Surveys in Geophysics*, 8, 25–83.

IBC (2009) *International Building Code*. International Code Council, Washington.

Idriss, I. M. (1990), "Response of Soft Soil Sites during Earthquakes", Proceedings, Memorial Symposium to honor Professor Harry Bolton Seed, Berkeley, California, Vol. II, May, pp 273–289.

Iyengar, R.N. & Ghosh, S. (2004) Microzonation of earthquake hazard in greater Delhi area. *Current Science*, 87, 1193–1202.

Iyengar, R.N.et al. (2010) Development of probabilistic seismic hazard map of India, technical report of the Working Committee of Experts (WCE) constituted by the National Disaster Management Authority, Govt. of India, New Delhi.

Jaiswal, K. & Sinha, R. (2007) Probabilistic seismic-hazard estimation for peninsular India. *Bulletin of the Seismological Society of America*, 97(1B), 318–330.

Johnston, A.C. (1996) Seismic moment assessment of earthquakes in stable continental regions, I: Instrumental seismicity. *Geophysics Journal International*, 124, 381–414.

Joshi, A., Kapil, M. & Patel, R.C. (2007) A deterministic approach for preparation of seismic hazard maps in North East India. *Natural Hazards*, 43, 129–146.

Joyner, W.B. & Boore, D.M. (1981) Peak horizontal acceleration and velocity from strong-motion records including records from the 1979 Imperial Valley. *California, Earthquake, Bulletin of the Seismological Society of America*, 71, 2011–2038.

Kaila, K.L. & Sarkar, D. (1978) Atlas of isoseismal maps of major earthquakes in India. *Geophysical Research Bulletin*, 16, 234–267.

Kalkan, E., Gülkan, P., Yilmaz, N. & Çelebi, M. (2009) Reassessment of probabilistic seismic hazard in the Marmara Region. *Bulletin of the Seismological Society of America*, 99(4), 2127–2146.

Kanamori, H. (1977). The energy release in great earthquakes. *Journal of Geophysical Research*, 82(20), 2981–2987.

Kayal, J.R. (2007) Recent large earthquakes in India: Seismotectonic Perspective. *International Association for Gondwana Research*, Japan, IAGR Memoir, 10, 189–199.

Kayal, J.R. (2008). Microearthquake seismology and seismotectonics of South Asia. *Springer Science & Business Media*, Heidelberg, Germany.

Kelkar, Y.N. (1968) Earthquakes experienced in Maharashtra during the last 300 years. *Daily Kesari Poona*, Marathi, January 7.

Khattri, K.N., Rogers, A.M., Perkins, D.M. & Algermissen, S.T. (1984) A seismic hazard map of India and adjacent areas. *Tectonophysics*, 108, 93–134.

Kijko, A. (2004) Estimation of the maximum earthquake magnitude, mmax. *Pure and Applied Geophysics*, 161, 1655–1681.

Kijko, A. & Sellevoll, M.A. (1989) Estimation of earthquake hazard parameters from incomplete data files, Part I: Utilization of extreme and complete catalogs with different threshold magnitudes. *Bulletin of the Seismological Society of America*, 79, 645–654.

Kiratzi, A.A., Karakaisis, G.F., Papadimitriou, E.E. & Papazachos, B.C. (1985) Seismic source-parameter relations for earthquakes in Greece. *Pure and Applied Geophysics*, 123, 27–41.

Kiureghian, D.A. & Ang, A.H.S. (1977) A fault-rupture model for seismic risk analysis. *Bulletin of the Seismological Society of America*, 67(4), 1173–1194.

Knopoff, L. (1964) The statistics of earthquakes in Southern California. *Bulletin of the Seismological Society of America*, 54(6), 1871–1873.

Kolathayar, S. & Sitharam, T.G. (2012) Comprehensive probabilistic seismic hazard analysis of Andaman-Nicobar regions. *Bulletin of Seismological Society of America*, 102(5), 2063–2076.

Kolathayar, S., Sitharam, T.G. & Vipin, K.S. (2012) Deterministic seismic hazard macrozonation of India. *Journal of Earth System Sciences*, Springer, 121(5), 1351–1364.

Koulakov, I., Bindi, D., Parolai, S., Grosser, H. & Milkereit, C. (2010). Distribution of seismic velocities and attenuation in the crust beneath the North Anatolian Fault (Turkey) from local earthquake tomography. *Bulletin of the Seismological Society of America*, 100(1), 207–224.

Kramer, S.L. (1996). Geotechnical earthquake engineering. *Prentice–Hall International Series in Civil Engineering and Engineering Mechanics*. Prentice-Hall, New Jersey.

Kramer, S.L. & Mayfield, R.T. (2007) Return period of soil liquefaction. *Journal of Geotechnical and Geoenvironmental Engineering*, 133(7), 802–813.

Krinitzsky, E.L. (2003) How to combine deterministic and probabilistic methods for assessing earthquake hazards. *Engineering Geology*, 70, 157–163.

Krinitzsky, E.L. (2005) Comment on J.U. Klugel's "Problems in the application of the SSHAC probability method for assessing earthquake hazards at Swiss nuclear power plants". *Engineering Geology*, 82, 66–68.

Krishnan, M.S. (1953) The structure and tectonic history of India. *Memoir of Geological Survey of India*, 81, 137.

Kumar, A., Anbazhagan, P. & Sitharam, T.G. (2013) Seismic hazard analysis of Lucknow considering local and active seismic gaps. *Natural Hazards*, 69(1), 327–350.

Kumar, P. (2011) *Seismic Microzonation of Imphal City and Probabilistic Seismic Hazard Assessment of Manipur State*. PhD Thesis, Indian Institute of Technology, Guwahati.

Kumar, P., Yuan, X., Ravi Kumar, M., Kind, R., Li, X. & Chadha, R.K. (2007) The rapid drift of Indian tectonic plate. *Nature*, 449, 894–897.

Kumar, S., Steven, G.W., Thomas, K.R., Daniel, R., Thakur, V.C. & Seitz, G.G. (2001) Earthquake recurrence and rupture dynamics of Himalayan frontal thrust, India. *Science*, 294, 2328–2331.

Lapajne, J., Motnikar, B.S. & Zupancic, P. (2003) Probabilistic seismic hazard assessment methodology for distributed seismicity. *Bulletin of the Seismological Society of America*, 93(6), 2502–2515.

Liao, S.S.C., Veneziano, D. & Whitman, R.V. (1988) Regression models for evaluating liquefaction probability. *Journal of Geotechnical Engineering*, 14(4), 389–411.

Lin, P.S. & Lee, C.T. (2008) Ground-motion attenuation relationships for subduction-zone earthquakes in Northeastern Taiwan. *Bulletin of the Seismological Society of America*, 98, 220–240.

Liu, R., Chen, Y., Ren, X., Xu, Z., Sun, L., Yang, H., Liang, J. & Ren, K. (2007) Comparison between different earthquake magnitudes determined by China Seismograph Network. *Acta Seismologica Sinica*, 20(5), 497–506.

Lunne, T., Robertson, P.K. & Powell, J.J.M. (1997). Cone penetration testing. *Geotechnical Practice*, 20, 23–35.

McGuire, R.M. & Arabasz, W.J. (1990) An introduction to probabilistic seismic hazard analysis. S.H. Ward ed. Geotechnical and environmental geophysics. *Society of Exploration Geophysicist*, 1, 333–353.

Mahajan, A.K., Thakur, V.C., Sharma, M.L. & Chauhan, M. (2010). Probabilistic seismic hazard map of NW Himalaya and its adjoining area, India. *Natural Hazards*, 53(3), 443–457.

Maiti, S.K., Nath, S.K., Adhikari, M.D., Srivastava, N., Sengupta, P. & Gupta, A.K. (2017) Probabilistic seismic hazard model of West Bengal, India. *Journal of Earthquake Engineering*, 21(7), 1113–1157.

Malagnini, L., Mayeda, K., Uhrhammer, R., Akinci, A. & Herrmann, R.B. (2007). A regional ground-motion excitation/attenuation model for the San Francisco region. *Bulletin of the Seismological Society of America*, 97(3), 843–862.

Mandal, P., Rastogi, B.K. & Sarma, C.S.P. (1998) Source parameter of Koyna earthquakes, India. *Bulletin of the Seismological Society of America*, 88, 833–842.

Martin, C., Secanell, R., Combes, P. & Lignon, G. (2002). Preliminary probabilistic seismic hazard assessment of France. *Proceedings of the 12th ECEE*, Paper Reference 870, September, London, England.

Matsuoka, M., Wakamatsu, K., Fujimoto, K. & Midorikawa, S. (2005) Nationwide site amplification zoning using GIS-based Japan engineering geomorphologic classification map. *Proc. 9th Int. Conf. on Struct. Safety and Reliability*, pp. 239–246.

McCann, W.R., Nishenko, S.P., Sykes, L.R. & Krause, J. (1979) Seismic gaps and plate tectonics: seismic potential for major boundaries. In: *Earthquake Prediction and Seismicity Patterns*. Birkhäuser, Basel. pp. 1082–1147.

Menon, A., Ornthammarath, T., Corigliano, M. & Lai, C.G. (2010) Probabilistic seismic hazard macrozonation of Tamil Nadu in Southern India. *Bulletin of the Seismological Society of America*, 100, 1320–1341.

Mohanty, W.K. & Walling, M.Y. (2008) First order seismic microzonation of Haldia, Bengal basin (India) using a GIS platform. *Pure and Applied Geophysics*, 165, 1325–1350.

Molchan, G. & Dmitrieva, O. (1992) Aftershock identification: Methods and new approaches. *Geophysical Journal International*, 109, 501–516.

Molnar, P. & Tapponnier, P. (1979) The collision between India and Eurasia. In: *Earthquakes and Volcanoes, Procs., from Scientific American*. WH Freeman and Company, San Francisco. pp. 62–73.

Mukhopadhyay, B. & Dasgupta, S. (2015) Seismic hazard assessment of Kashmir and Kangra valley region, Western Himalaya, India. *Geomatics, Natural Hazards and Risk*, 6(2), 149–183.

Mukhopadhyay, M. (1984) Seismotectonics of subduction and backarc rifting under the Andaman Sea. *Tectonophysics*, 108, 229–239.

Mukhopadhyay, M. (1988) Gravity anomalies and deep structure of the Andaman arc. *Marine Geophysical Researches*, 9(3), 197–210.

Muthuganeisan, P. &Raghu Kanth, S.T.G. (2016) Site-specific probabilistic seismic hazard map of Himachal Pradesh, India, Part I: Site-specific ground motion relations. *Acta Geophysica*, 64(2), 336–361.

Naik, N. & Choudhury, D. (2015) Deterministic seismic hazard analysis considering different seismicity levels for the state of Goa, India. *Natural Hazards*, 75(1), 557–580.

Nakata, T. (1989) Active faults of the Himalaya of India and Nepal. *Geological Society of America Special Paper*, 232, 243–264.

Nath, S.K. (2006) Seismic hazard and microzonation atlas of the Sikkim Himalaya. In: *Published by Department of Science and Technology*. Government of India, New Delhi.

Nath, S.K., Shukla, K. & Vyas, M. (2008) Seismic hazard scenario and attenuation model of the Garhwal Himalaya using near-field synthesis from weak motion seismometry. *Journal of Earth System Science*, 117(2), 649–670.

Nath, S.K. (2011) *Seismic Microzonation Handbook*. Geoscience Division, Ministry of Earth Science, Govt. of India, New Delhi.

Nath, S.K., Thingbaijam, K.K.S. & Raj, A. (2008) Earthquake hazard in northeast India—A seismic microzonation approach with typical case studies from Sikkim Himalaya and Guwahati city. *Journal of Earth System Science*, 117(2), 809–831.

Nath, S.K., Thingbaijam, K.K.S., Adhikari, M.D., Nayak, A., Devaraj, N., Ghosh, S.K. & Mahajan, A.K. (2013) Topographic gradient based site characterization in India complemented by strong ground-motion spectral attributes. *Soil Dynamics and Earthquake Engineering*, 55, 233–246.

Nath, S.K., Adhikari, M.D., Maiti, S.K., Devaraj, N., Srivastava, N. & Mohapatra, L.D. (2014) Earthquake scenario in West Bengal with emphasis on seismic hazard microzonation of the city of Kolkata, India. *Natural Hazards and Earth System Sciences*, 14(9), 2549.

Nath, S.K., Raj, A., Thingbaijam, K.K.S. & Kumar, A. (2009) Ground motion synthesis and seismic scenario in Guwahati city-a stochastic approach. *Seismological Research Letters*, 80, 233–242.

Nath, S.K. & Thingbaijam, K.K.S. (2010) Peak ground motion predictions in India: An appraisal for rock sites. *Journal of Seismology*, 15(2), 295–315.

Nath, S.K. & Thingbaijam, K.K.S. (2012) Probabilistic seismic hazard assessment of India. *Seismological Research Letters*, 83(1), 135–149.

Nath, S.K., Vyas, M., Pal, I. & Sengupta, P. (2005) A hazard scenario in the Sikkim Himalaya from seismotectonics spectral amplification source parameterization and spectral attenuation laws using strong motion seismometry. *Journal of Geophysical Research*, 110, B01301. DOI: 10.1029/2004/2004JB003199.

Nath, S.K., Vyas, M., Pal, I., Singh, A.K., Mukherjee, S. & Sengupta, P. (2006) Spectral attenuation models in the Sikkim Himalaya from the observed and simulated strong motion events in the region. *Current Science*, 88(2), 295–303.

Nayak, M., Sitharam, T.G. & Kolathayar, S. (2015) A revisit to seismic hazard at Uttarakhand. *International Journal of Geotechnical Earthquake Engineering (IJGEE)*, 6(2), 56–73.

Ni, J. & Barazangi, M. (1986) Seismotectonics of the Zagros continental collision zone and a comparison with the Himalayas. *Journal of Geophysical Research*, 91, 8205–8218.

Nuttli, O.W. (1983) Average seismic source parameter relations for mid-plate earthquakes. *Bulletin of the Seismological Society of America*, 73, 519–535.

Oldham, T. (1883) A catalogue of Indian earthquakes from the earliest time to the end of A.D. 1869. *Memoirs of the Geological Survey of India*, 29, 163–215.

Ordaz, M., Aguilar, A. & Arboleda, J. (2007) CRISIS2007 – Ver. 1.1: Program for computing seismic hazard. In: *Instituto de Ingenieria*, UNAM, Mexico.

Pallav, K., Raghu Kanth, S.T.G. & Singh, K.D. (2012) Probabilistic seismic hazard estimation of Manipur, India. *Journal of Geophysics and Engineering*, 9(5), 516.

Pallav, K., Raghukanth, S.T.G. & Singh, K.D. (2015). Estimation of Seismic Site Coefficient and Seismic Microzonation of Imphal City, India, Using the Probabilistic Approach. *Acta Geophysica*, 63(5), 1339–1367.

Panza, G.F., Vaccari, F. & Cazzaro, R. (1999). Deterministic seismic hazard assessment. In Vrancea earthquakes: tectonics, hazard and risk mitigation (pp. 269-286). Springer, Dordrecht.

Papazachos, B.C., Karakostas, V.G., Kiratzi, A.A., Margaris, B.N., Papazachos, C.B. & Scordilis, E.M. (2002) Uncertainties in the estimation of earthquake magnitudes in Greece. *Journal of Seismology*, 6, 557–570.

Parvez, I.A., Vaccari, F. & Panza, G.F. (2003) A deterministic seismic hazard map of India and adjacent areas. *Geophysics Journal International*, 155, 489–508.

Patro, B.P.K., Nandini, N. & Sarma, S.V.S. (2006) Crustal geoelectric structure and the focal depths of major stable continental region earthquakes in India. *Current Science*, 90(1), 10, 107–113.

Patton, H.J. & Walter, W.R. (1993) Regional moment: Magnitude relations for earthquakes and explosions. *Geophysical Research Letters*, 20(4), 277–280.

Puri, N. & Jain, A. (2016) Deterministic seismic hazard analysis for the state of Haryana, India. *Indian Geotechnical Journal*, 46(2), 164–174.

Purnachandra Rao, N. (1999) Single station moment tensor inversion for focal mechanisms of Indian intra-plate earthquakes. *Current Science*, 77, 1184–1189.

Raghu Kanth, S.T.G. (2010) Estimation of seismicity parameters for India. *Seismological Research Letters*, 81(2), 207–217.

Raghu Kanth, S.T.G. & Iyengar, R.N. (2006) Seismic hazard estimation for Mumbai city. *Current Science*, 91(11), 1486–1494.

Raghu Kanth, S.T.G. & Iyengar, R.N. (2007) Estimation of seismic spectral acceleration in Peninsular India. *Journal of Earth System Sciences*, 116(3), 199–214.

Rai, D.C. & Murty, C.V.R. (2003) Reconnaissance report: North Andaman (Diglipur) earthquake of 14 September 2002, Dept. of Science and Technology, Government of India, New Delhi, United Nations Development Programme, New Delhi, India, and NICEE, IIT Kanpur, India.

Rajendran, K. & Gupta, H.K. (1989) Seismicity and tectonic stress field of a part of the Burma-Andaman-Nicobar Arc. *Bulletin of the Seismological Society of America*, 79, 989–1005.

Rajendran, K. & Rajendran, C.P. (2011) Revisiting some significant earthquake sources in the Himalaya: Perspectives on past seismicity. *Tectonophysics*, 504, 75–88.

Rao, B.R. & Rao, P.S. (1984) Historical seismicity of peninsular India. *Bulletin of the Seismological Society of America*, 74(6), 2519–2533.

Rao, B.V. & Murty, B.V.S. (1970) Earthquakes and tectonics in peninsular India. *Journal of Indian Geophysical Union*, 7, 1–8.

Rastogi, B.K. (1974) Earthquake mechanisms and plate tectonics in the Himalayan region. *Tectonophysics*, 21, 47–56.

Ray, J.S. (2006) Age of the Vindhyan supergroup: A review of recent findings. *Journal of Earth System Sciences*, 115(1), 149–160.

Reasenberg, P. (1985) Second-order moment of central California seismicity 1969–82. *Journal of Geophysical Research*, 90, 5479–5495.

Reddy, P.R. (2003) Need for high-resolution deep seismic reflection studies in strategic locales of South India. *Current Science*, 84(8), 973–974.

Richter, C.F. (1935) An instrumental earthquake magnitude scale. *Bulletin of the Seismological Society of America*, 25, 1–32.

Rodriguez-Marek, A., Bray, J.D. & Abrahamson, N.A. (2001) An empirical geotechnical seismic site response procedure. *Earthquake Spectra*, 17(1), 65–87.

Rydelek, P.A. & Sacks, I.S. (1989) Testing the completeness of earthquake catalogs and the hypothesis of self-similarity. *Nature*, 337, 251–253.

Savage, W.U. (1972) Microearthquake clustering near Fairview Peak Nevada, and in the Nevada Seismic Zone. *Journal of Geophysical Research*, 77(35), 7049–7056.

Scordilis, E.M. (2006) Empirical global relations converting Ms and m_b to moment magnitude. *Journal of Seismology*, 10, 225–236.

Seeber, L., Armbruster, J.G. & Jacob, K.H. (1999) Probabilistic assessment of earthquake hazard for the state of Maharashtra. Report to Government of Maharashtra Earthquake Rehabilitation Cell, Mumbai.

Seed, H.B. & Idriss, I.M. (1971) Simplified procedure for evaluating soil liquefaction potential. *Journal of Soil Mechanics and Foundation*, 97, 1249–1273.

Seed, H.B. & Idriss, I.M. (1982) Ground motions and soil liquefaction during earthquakes. *Earthquake Engineering Research Institute*, Monograph Series, 5.

Seed, H.B., Tokimatsu, K., Harder, L.F. & Chung, R.M. (1985) Influence of SPT procedures in soil liquefaction resistance evaluations. *Journal of Geotechnical Engineering*, 111(12), 1425–1445.

Shanker, D. & Sharma, M.L. (1998) Estimation of seismic hazard parameters for the Himalayas and its vicinity from complete data files. *Pure and Applied Geophysics*, 152, 267–279.

Sharma, M.L. (1998) Attenuation relationship for estimation of peak ground acceleration using data from strong-motion arrays in India. *Bulletin of the Seismological Society of America*, 88, 1063–1069.

Sharma, B., Gupta, A.K., Devi, D.K., Kumar, D., Teotia, S.S. & Rastogi, B.K. (2008). Attenuation of high-frequency seismic waves in Kachchh Region, Gujarat, India. *Bulletin of the Seismological Society of America*, 98(5), 2325–2340.

Sharma, M.L., Douglas, J., Bungum, H. & Kotadia, J. (2009) Ground-motion prediction equations based on data from the Himalayan and Zagros regions. *Journal of Earthquake Engineering*, 13, 1191–1210.

Sharma, M.L. & Malik, S. (2006) Probabilistic seismic hazard analysis and estimation of spectral strong ground motion on Bed rock in North East India, 4th. *Int. Conf. Earthquake Engineering, Taipei*, Taiwan, Oct 12–13, 2006.

Sil, A., Sitharam, T.G. & Kolathayar, S. (2013) Probabilistic seismic hazard analysis of Tripura and Mizoram states. *Natural Hazards*, 68(2), 1089–1108.

Singh, R.K. (2009) *Probabilistic Seismic Hazard and Risk Analysis: A Case Study for Ahmedabad City*. PhD Thesis, Indian Institute of Technology, Kanpur.

Sitharam, T.G. & Anbazhagan, P. (2007) Seismic hazard analysis for the Bangalore region. *Natural Hazards*, 40, 261–278.

Sitharam, T.G. & Anbazhagan, P. (2008) Seismic microzonation: Principles, practices and experiments. *EJGE Special Volume Bouquet*, 8, www.ejge.com/Bouquet08/Sitharam/Sitharam_ppr.pdf.

Sitharam, T.G. & Kolathayar, S. (2013) Seismic hazard analysis of India using areal sources. *Journal of Asian Earth Sciences (Elsevier)*, 62, 647–653.

Sitharam, T.G., Kolathayar, S. & James, N. (2015) Probabilistic assessment of surface level seismic hazard in India using topographic gradient as a proxy for site condition. *Geoscience Frontiers*, 6(6), 847–859.

Sitharam, T.G. & Vipin, K.S. (2011) Evaluation of spatial variation of peak horizontal acceleration and spectral acceleration for South India: A probabilistic approach. *Natural Hazards*, 59(2), 639–653.

Soghrat, M.R. & Ziyaeifar, M. (2017) Ground motion prediction equations for horizontal and vertical components of acceleration in Northern Iran. *Journal of Seismology*, 21(1), 99–125.

Srivastava, H.N. & Ramachandran, K. (1985) New catalog of earthquakes for peninsular India during 1839–1900. *Mausam*, 36(3), 351–358.

Stein, R.S. & Hanks, T.C. (1998) M ≥ 6 earthquakes in Southern California during the twentieth century: No evidence for a seismicity or moment deficit. *Bulletin of the Seismological Society of America*, 88, 635–652.

Stepp, J.C. (1972, November). Analysis of completeness of the earthquake sample in the Puget Sound area and its effect on statistical estimates of earthquake hazard. *Proc. of the 1st Int. Conf. on Microzonazion*, Seattle (Vol. 2, pp. 897–910).

Stepp, J.C., Wong, I., Whitney, J., Quittemeyer, R., Abrahamson, N., Toro, G., Youngs, R., Coppersmith, K., Savy, J. & Sullivan, T. (2001) Yucca mountain PSHA project members, probabilistic seismic hazard analyses for ground motions and fault displacements at Yucca Mountain, Nevada. *Earthquake Spectra*, 17, 113–151.

Stiphout, V.T., Zhuang, J. & Marsan, D. (2010) Seismicity declustering, community online resource for statistical seismicity analysis. DOI: 10.3929/ethz, www.corssa.org.

Tandon, A.N. & Srivastava, H.N. (1974) Earthquake occurrence in India. In *Earthquake Engineering – Jai Krishna Commemoration Volume*. Sarita Prakashan, Meerut, India.

Thingbaijam, K.K.S., Nath, S.K., Yadav, A., Raj, A., Walling, M.Y. & Mohanty, W.K. (2008) Recent seismicity in Northeast India and its adjoining region. *Journal of Seismology*, 12, 107–123.

Uhrhammer, R.A. (1986) Characteristic of northern and central California seismicity abstract. *Earthquake Notes*, 1, 21.

US Nuclear Regulatory Commission. (1997) Identification and characterization of seismic sources and determination of safe shutdown earthquake ground motion. *Regulatory Guide*, 1, 01–165.

USCOLD: 1995, Guidelines for Earthquake Design and Evaluation of Structures Appurtenant to Dams, *United States Committee on Large Dams*. 75 p

Utsu, T. (1965) A method for determining the value of b in the formula log N=a-bM showing the magnitude – frequency relation for the earthquakes. *Geophysics Bulletin, Hokkaido University*, 13, 99–103.

Utsu, T. (1999) Representation and analysis of the earthquake size distribution: A historical review and some new approaches. *Pure and Applied Geophysics*, 155, 509–535.

Valdiya, K.S. (1992) The main boundary thrust zone of the Himalaya, India. *Annales Tectonicae*, 54–84, Supplement to volume 6.

Vilanova, S.P. & Fonseca, J.F.B.D. (2007) Probabilistic seismic-hazard assessment for Portugal. *Bulletin of the Seismological Society of America*, 93(6), 2502–2515.

Vipin, K.S., Anbazhagan, P. & Sitharam, T.G. (2009) Estimation of peak ground acceleration and spectral acceleration for South India with local site effects: Probabilistic approach. *Natural Hazards Earth System Science*, 9, 865–878.

Vipin, K.S. & Sitharam, T.G. (2011). Multiple source and attenuation relationships for evaluation of deterministic seismic hazard: Logic tree approach considering local site effects. Georisk: *Assessment and Management of Risk for Engineered Systems and Geohazards*, 5(3-4), 173–185.

Vipin, K.S., Sitharam, T.G. & Kolathayar, S. (2013) Assessment of seismic hazard and liquefaction potential of Gujarat based on probabilistic approaches. *Natural Hazards*, 65(2), 1179–1195.

Wahlstrom, E. & Grunthal, R. (2000) Probabilistic seismic hazard assessment (horizontal PGA) for Sweden, Finland and Denmark using different logic tree approaches. *Soil Dynamics and Earthquake Engineering*, 20, 45–58.

Wald, D.J. & Allen, T.I. (2007) Topographic slope as a proxy for seismic site conditions and amplification. *Bulletin of the Seismological Society of America*, 97(5), 1379–1395.

Wells, D.L. & Coppersmith, K.J. (1994) New empirical relationships among magnitude, rupture length, rupture width, rupture area, and surface displacement. *Bulletin of the Seismological Society of America*, 84, 974–1002.

Wiemer, S. (2001) A software package to analyze seismicity: Zmap. *Seismological Research Letters*, 72(2), 374–383.

Wiemer, S. & Wyss, M. (1994) Seismic quiescence before the Landers (M=7.5) and Big Bear (M=6.5) 1992 earthquakes. *Bulletin of the Seismological Society of America*, 84, 900–916.

Wiemer, S. & Wyss, M. (1997) Mapping the frequency-magnitude distribution in asperities: An improved technique to calculate recurrence times? *Journal of Geophysical Research*, 102, 15115–15128.

Wiemer, S. & Wyss, M. (2000) Minimum magnitude of complete reporting in earthquake catalogs: Examples from Alaska, the Western United States, and Japan. *Bulletin of the Seismological Society of America*, 90, 859–869.

Wiemer, S. & Wyss, M. (2002) Mapping spatial variability of the frequency-magnitude distribution of earthquakes. *Advances in Geophysics*, 45, 259–302.

Wills, C.J., Petersen, M.D., Bryant, W.A., Reichle, M.S., Saucedo, G.J., Tan, S.S., Taylor, G.C. & Treiman, J.A. (2000) A site-conditions map for California based on geology and shear wave velocity. *Bulletin of the Seismological Society of America*, 90, S187–S208.

Woessner, J. & Wiemer, S. (2005) Assessing the quality of earthquake catalogs: Estimating the magnitude of completeness and its uncertainty. *Bulletin of the Seismological Society of America*, 95(2), 684–698.

Woo, G. (1996) Kernel estimation methods for seismic hazard area source model. *Bulletin of Seismological Society of America*, 86(2), 253–362.

Youd, T.L. & Idriss, I.M. (2001). Liquefaction resistance of soils: summary report from the 1996 NCEER and 1998 NCEER/NSF workshops on evaluation of liquefaction resistance of soils. *Journal of geotechnical and geoenvironmental engineering*, 127(4), 297–313.

Youngs, R.R., Chiou, S.J., Silva, W.J. & Humphrey, J.R. (1997) Strong ground motion relationships for subduction earthquakes. *Seismological Research Letters*, 68, 58–73.

Yu, Y.X. & Wang, S.Y. (2004). Attenuation relations for horizontal peak ground acceleration and response spectrum in northeastern Tibetan Plateau region. *Acta Seismologica Sinica*, 17(6), 651–661.

Zhao, J.X., Jiang, F., Shi, P., Xing, H., Huang, H., Hou, R. & Somerville, P.G. (2016) Ground-motion prediction equations for subduction slab earthquakes in Japan using site class and simple geometric attenuation functions. *Bulletin of the Seismological Society of America*, 106(4), 1535–1551.

Zhao, J.X., Zhang, J., Asano, A., Ohno, Y., Oouchi, T., Takahashi, T., Ogawa, H., Irikura, K., Thio, H.K., Somerville, P.G., Fukushima, Y. & Fukushima, Y. (2006) Attenuation relations of strong ground motion in Japan using site classification based on predominant period. *Bulletin of the Seismological Society of America*, 96, 898–913.

Appendix A: Magnitude conversion relations

Magnitude conversion relations were developed using data from active tectonic regions and stable continental regions separately (Figures A.1, A.2, A.3, and A.4).

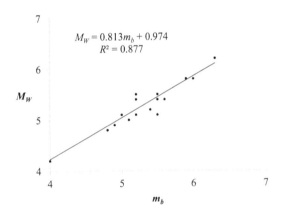

Figure A.1 Relation between m_b and M_W developed from events of shield region

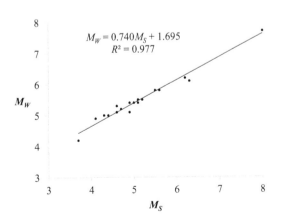

Figure A.2 Relation between M_S and M_W developed from events of shield region

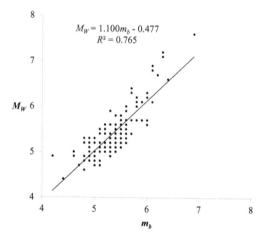

Figure A.3 Relation between m_b and M_W developed from events of active tectonic regions

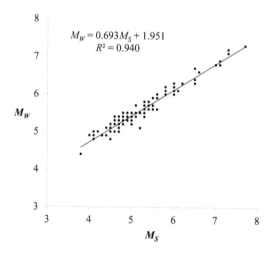

Figure A.4 Relation between M_s and M_W developed from events of active tectonic regions

Table A.1 Comparison of magnitude conversion relations relating m_b with M_W

m_b	M_W		
	Relation from entire data set	Relation from data in stable continental region	Relation from data in active tectonic region
3.5	3.9	3.8	3.4
4.0	4.4	4.2	3.9
4.5	4.8	4.6	4.5
5.0	5.3	5.0	5.0
5.5	5.7	5.4	5.6
6.0	6.2	5.9	6.1

Table A.2 Comparison of magnitude conversion relations relating M_s with M_w

M_s	M_w		
	Relation from entire data set	Relation from data in stable continental region	Relation from data in active tectonic region
3.0	4.0	3.9	4.0
4.0	4.7	4.7	4.7
5.0	5.4	5.4	5.4
6.0	6.1	6.1	6.1
7.0	6.8	6.9	6.8
8.0	7.5	7.6	7.5

Tables A.1 and A.2 provide the comparison of M_w obtained from these relations with that obtained using the entire data set (as presented in Chapter 2).

Appendix B: Frequency magnitude distribution plots of seismic source zones

This appendix presents the frequency magnitude distribution (FMD) plots of regional seismic source zones identified in India and adjoining areas (Chapter 4). FMDs were not plotted for three source zones (95, 98, and 99) where seismic activity is very low (no events reported with $M_W \geq 4$), and those zones are considered to be inactive.

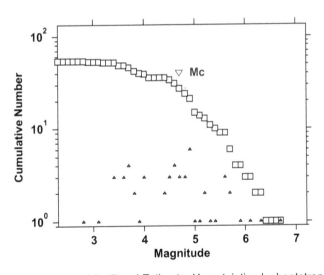

Maximum Likelihood Estimate, Uncertainties by bootstrapping
b-value = 0.82 +/- 0.27, a value = 5.38, a value (annual) = 3.58
Magnitude of Completeness = 4.7 +/- 0.42

Figure B.1 Frequency magnitude distribution for source zone 1

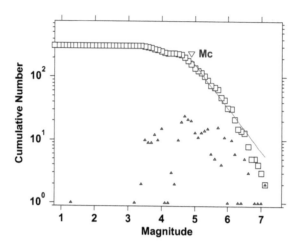

Maximum Likelihood Estimate, Uncertainties by bootstrapping
b-value = 0.69 +/- 0.18, a value = 5.65, a value (annual) = 3.73
Magnitude of Completeness = 4.9 +/- 0.36

Figure B.2 Frequency magnitude distribution for source zone 2

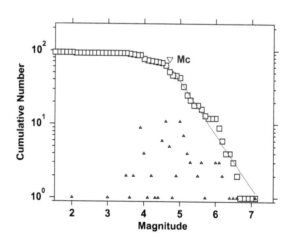

Maximum Likelihood Estimate, Uncertainties by bootstrapping
b-value = 0.72 +/- 0.16, a value = 5.2, a value (annual) = 3.32
Magnitude of Completeness = 4.7 +/- 0.35

Figure B.3 Frequency magnitude distribution for source zone 3

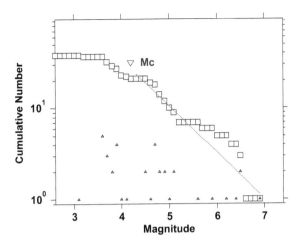

Maximum Likelihood Estimate, Uncertainties by bootstrapping
b-value = 0.51 +/- 0.07, a value = 3.55, a value (annual) = 1.81
Magnitude of Completeness = 4.2 +/- 0.72

Figure B.4 Frequency magnitude distribution for source zone 4

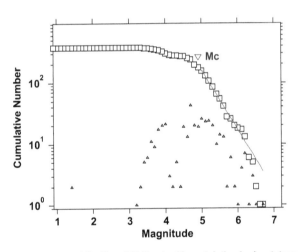

Maximum Likelihood Estimate, Uncertainties by bootstrapping
b-value = 0.97 +/- 0.14, a value = 7.05, a value (annual) = 5.15
Magnitude of Completeness = 4.9 +/- 0.18

Figure B.5 Frequency magnitude distribution for source zone 5

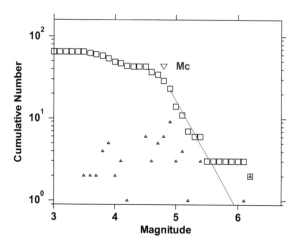

Maximum Likelihood Estimate, Uncertainties by bootstrapping
b-value = 1.34 +/- 0.35, a value = 7.9, a value (annual) = 6.07
Magnitude of Completeness = 4.8 +/- 0.21

Figure B.6 Frequency magnitude distribution for source zone 6

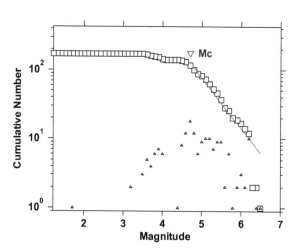

Maximum Likelihood Estimate, Uncertainties by bootstrapping
b-value = 0.73 +/- 0.11, a value = 5.53, a value (annual) = 3.59
Magnitude of Completeness = 4.7 +/- 0.18

Figure B.7 Frequency magnitude distribution for source zone 7

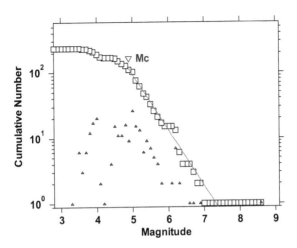

Maximum Likelihood Estimate, Uncertainties by bootstrapping
b-value = 0.85 +/- 0.13, a value = 6.22, a value (annual) = 4.27
Magnitude of Completeness = 4.9 +/- 0.19

Figure B.8 Frequency magnitude distribution for source zone 8

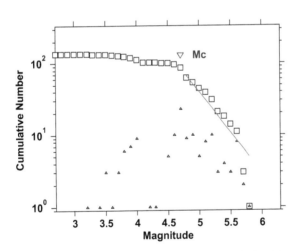

Maximum Likelihood Estimate, Uncertainties by bootstrapping
b-value = 1.13 +/- 0.15, a value = 7.23, a value (annual) = 5.61
Magnitude of Completeness = 4.7 +/- 0.07

Figure B.9 Frequency magnitude distribution for source zone 9

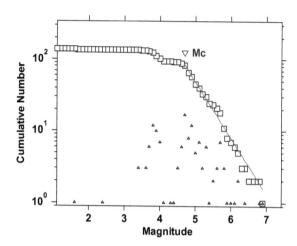

Maximum Likelihood Estimate, Uncertainties by bootstrapping
b-value = 0.79 +/- 0.16, a value = 5.62, a value (annual) = 3.82
Magnitude of Completeness = 4.7 +/- 0.26

Figure B.10 Frequency magnitude distribution for source zone 10

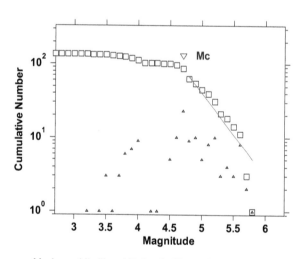

Maximum Likelihood Estimate, Uncertainties by bootstrapping
b-value = 1.13 +/- 0.15, a value = 7.23, a value (annual) = 5.61
Magnitude of Completeness = 4.7 +/- 0.07

Figure B.11 Frequency magnitude distribution for source zone 11

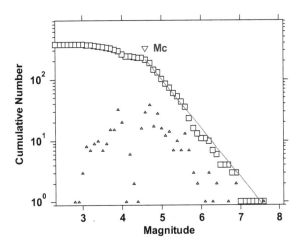

Maximum Likelihood Estimate, Uncertainties by bootstrapping
b-value = 0.8 +/- 0.09, a value = 5.99, a value (annual) = 4.05
Magnitude of Completeness = 4.6 +/- 0.12

Figure B.12 Frequency magnitude distribution for source zone 12

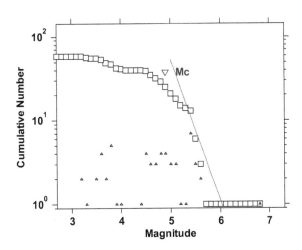

Maximum Likelihood Estimate, Uncertainties by bootstrapping
b-value = 1.68 +/- 1.21, a value = 10.1, a value (annual) = 8.29
Magnitude of Completeness = 4.9 +/- 0.65

Figure B.13 Frequency magnitude distribution for source zone 13

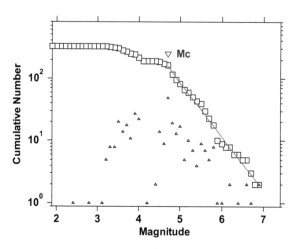

Maximum Likelihood Estimate, Uncertainties by bootstrapping
b-value = 0.86 +/- 0.14, a value = 6.24, a value (annual) = 4.07
Magnitude of Completeness = 4.7 +/- 0.17

Figure B.14 Frequency magnitude distribution for source zone 14

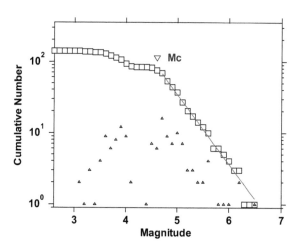

Maximum Likelihood Estimate, Uncertainties by bootstrapping
b-value = 0.98 +/- 0.25, a value = 6.42, a value (annual) = 4.5
Magnitude of Completeness = 4.6 +/- 0.27

Figure B.15 Frequency magnitude distribution for source zone 15

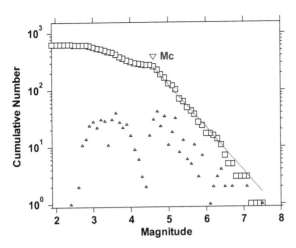

Maximum Likelihood Estimate, Uncertainties by bootstrapping
b-value = 0.76 +/- 0.07, a value = 5.9, a value (annual) = 3.21
Magnitude of Completeness = 4.6 +/- 0.17

Figure B.16 Frequency magnitude distribution for source zone 16

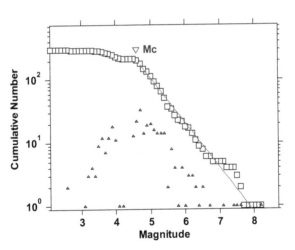

Maximum Likelihood Estimate, Uncertainties by bootstrapping
b-value = 0.71 +/- 0.07, a value = 5.57, a value (annual) = 3.58
Magnitude of Completeness = 4.6 +/- 0.11

Figure B.17 Frequency magnitude distribution for source zone 17

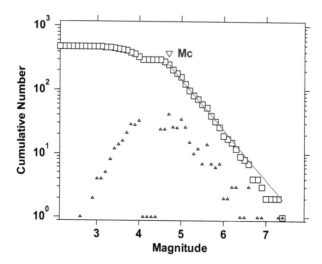

Maximum Likelihood Estimate, Uncertainties by bootstrapping
b-value = 0.79 +/- 0.08, a value = 6.09, a value (annual) = 3.84
Magnitude of Completeness = 4.7 +/- 0.16

Figure B.18 Frequency magnitude distribution for source zone 18

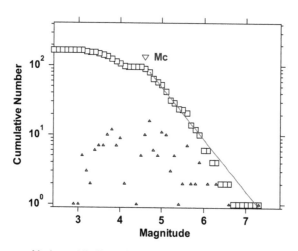

Maximum Likelihood Estimate, Uncertainties by bootstrapping
b-value = 0.75 +/- 0.17, a value = 5.44, a value (annual) = 3.34
Magnitude of Completeness = 4.6 +/- 0.38

Figure B.19 Frequency magnitude distribution for source zone 19

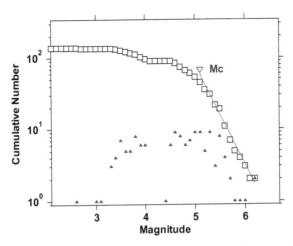

Maximum Likelihood Estimate, Uncertainties by bootstrapping
b-value = 1.39 +/- 0.51, a value = 8.85, a value (annual) = 6.95
Magnitude of Completeness = 5.1 +/- 0.41

Figure B.20 Frequency magnitude distribution for source zone 20

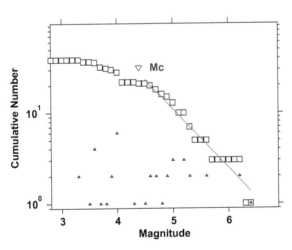

Maximum Likelihood Estimate, Uncertainties by bootstrapping
b-value = 0.65 +/- 0.23, a value = 4.29, a value (annual) = 2.39
Magnitude of Completeness = 4.4 +/- 0.71

Figure B.21 Frequency magnitude distribution for source zone 21

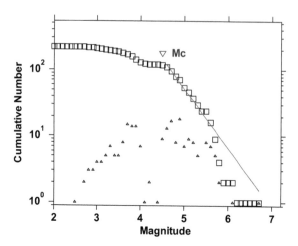

Maximum Likelihood Estimate, Uncertainties by bootstrapping
b-value = 0.9 +/- 0.28, a value = 6.2, a value (annual) = 4.29
Magnitude of Completeness = 4.5 +/- 0.45

Figure B.22 Frequency magnitude distribution for source zone 22

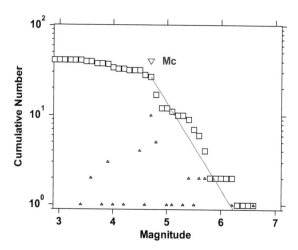

Maximum Likelihood Estimate, Uncertainties by bootstrapping
b-value = 0.96 +/- 0.22, a value = 5.96, a value (annual) = 3.76
Magnitude of Completeness = 4.7 +/- 0.13

Figure B.23 Frequency magnitude distribution for source zone 23

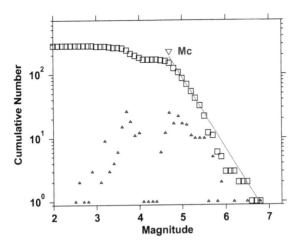

Maximum Likelihood Estimate, Uncertainties by bootstrapping
b-value = 1.11 +/- 0.47, a value = 7.48, a value (annual) = 5.33
Magnitude of Completeness = 4.7 +/- 0.5

Figure B.24 Frequency magnitude distribution for source zone 24

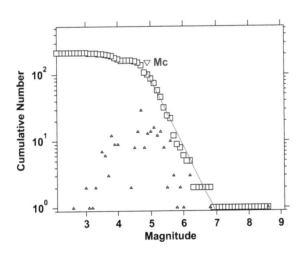

Maximum Likelihood Estimate, Uncertainties by bootstrapping
b-value = 1.01 +/- 0.26, a value = 6.99, a value (annual) = 5.11
Magnitude of Completeness = 4.9 +/- 0.2

Figure B.25 Frequency magnitude distribution for source zone 25

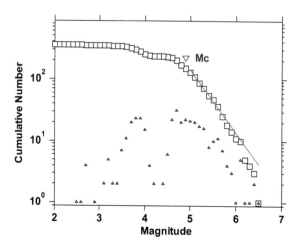

Maximum Likelihood Estimate, Uncertainties by bootstrapping
b-value = 1.02 +/- 0.32, a value = 7.25, a value (annual) = 5.31
Magnitude of Completeness = 4.9 +/- 0.37

Figure B.26 Frequency magnitude distribution for source zone 26

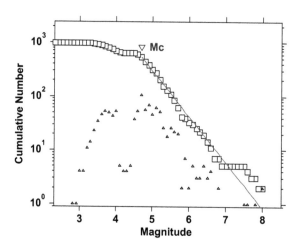

Maximum Likelihood Estimate, Uncertainties by bootstrapping
b-value = 0.85 +/- 0.07, a value = 6.73, a value (annual) = 4.44
Magnitude of Completeness = 4.7 +/- 0.09

Figure B.27 Frequency magnitude distribution for source zone 27

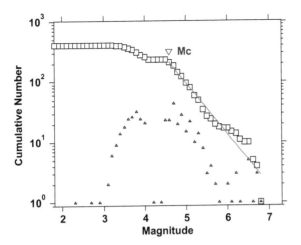

Maximum Likelihood Estimate, Uncertainties by bootstrapping
b-value = 0.85 +/- 0.14, a value = 6.22, a value (annual) = 4.34
Magnitude of Completeness = 4.6 +/- 0.22

Figure B.28 Frequency magnitude distribution for source zone 28

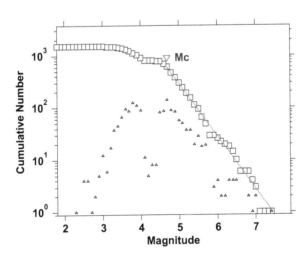

Maximum Likelihood Estimate, Uncertainties by bootstrapping
b-value = 1.01 +/- 0.07, a value = 7.52, a value (annual) = 5.46
Magnitude of Completeness = 4.7 +/- 0.06

Figure B.29 Frequency magnitude distribution for source zone 29

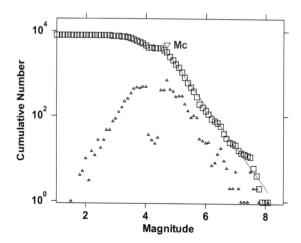

Maximum Likelihood Estimate, Uncertainties by bootstrapping
b-value = 0.97 +/- 0.08, a value = 8.04, a value (annual) = 5.79
Magnitude of Completeness = 4.7 +/- 0.16

Figure B.30 Frequency magnitude distribution for source zone 30

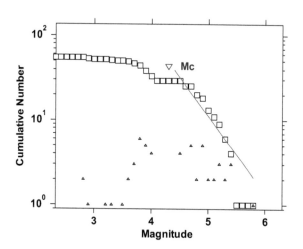

Maximum Likelihood Estimate, Uncertainties by bootstrapping
b-value = 0.91 +/- 0.44, a value = 5.58, a value (annual) = 3.71
Magnitude of Completeness = 4.3 +/- 0.51

Figure B.31 Frequency magnitude distribution for source zone 31

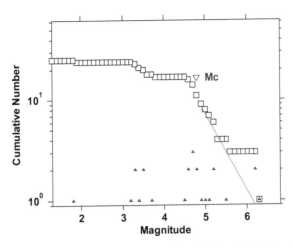

Maximum Likelihood Estimate, Uncertainties by bootstrapping
b-value = 0.76 +/- 0, a value = 4.67, a value (annual) = 2.53
Magnitude of Completeness = 4.8 +/- 0.85

Figure B.32 Frequency magnitude distribution for source zone 32

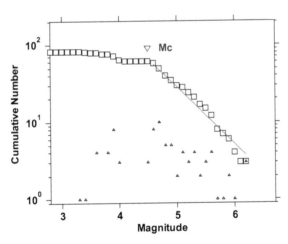

Maximum Likelihood Estimate, Uncertainties by bootstrapping
b-value = 0.74 +/- 0.2, a value = 5.14, a value (annual) = 3.31
Magnitude of Completeness = 4.5 +/- 0.36

Figure B.33 Frequency magnitude distribution for source zone 33

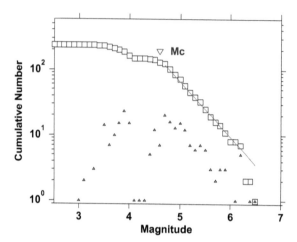

Maximum Likelihood Estimate, Uncertainties by bootstrapping
b-value = 0.84 +/- 0.2, a value = 6.02, a value (annual) = 4.09
Magnitude of Completeness = 4.6 +/- 0.33

Figure B.34 Frequency magnitude distribution for source zone 34

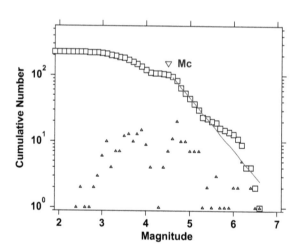

Maximum Likelihood Estimate, Uncertainties by bootstrapping
b-value = 0.79 +/- 0.17, a value = 5.59, a value (annual) = 3.6
Magnitude of Completeness = 4.5 +/- 0.35

Figure B.35 Frequency magnitude distribution for source zone 35

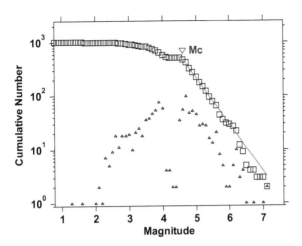

Maximum Likelihood Estimate, Uncertainties by bootstrapping
b-value = 0.87 +/- 0.07, a value = 6.7, a value (annual) = 4.04
Magnitude of Completeness = 4.6 +/- 0.07

Figure B.36 Frequency magnitude distribution for source zone 36

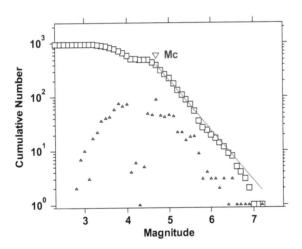

Maximum Likelihood Estimate, Uncertainties by bootstrapping
b-value = 0.93 +/- 0.08, a value = 6.97, a value (annual) = 4.82
Magnitude of Completeness = 4.7 +/- 0.12

Figure B.37 Frequency magnitude distribution for source zone 37

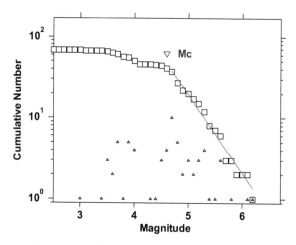

Maximum Likelihood Estimate, Uncertainties by bootstrapping
b-value = 0.97 +/- 0.29, a value = 6.13, a value (annual) = 4.48
Magnitude of Completeness = 4.6 +/- 0.37

Figure B.38 Frequency magnitude distribution for source zone 38

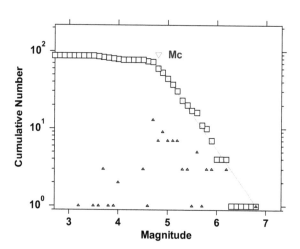

Maximum Likelihood Estimate, Uncertainties by bootstrapping
b-value = 0.9 +/- 0.14, a value = 6.16, a value (annual) = 4.33
Magnitude of Completeness = 4.8 +/- 0.16

Figure B.39 Frequency magnitude distribution for source zone 39

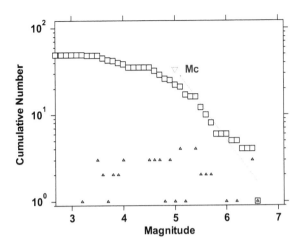

Maximum Likelihood Estimate, Uncertainties by bootstrapping
b-value = 0.82 +/- 0.56, a value = 5.64, a value (annual) = 3.95
Magnitude of Completeness = 5 +/- 0.74

Figure B.40 Frequency magnitude distribution for source zone 40

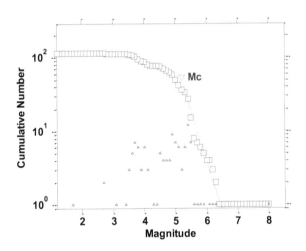

Maximum Likelihood Estimate, Uncertainties by bootstrapping
b-value = 1.34 +/- 0.53, a value = 8.63, a value (annual) = 6.38
Magnitude of Completeness = 5.2 +/- 0.38

Figure B.41 Frequency magnitude distribution for source zone 41

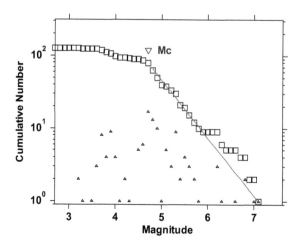

Maximum Likelihood Estimate, Uncertainties by bootstrapping
b-value = 0.8 +/- 0.16, a value = 5.64, a value (annual) = 3.38
Magnitude of Completeness = 4.7 +/- 0.12

Figure B.42 Frequency magnitude distribution for source zone 42

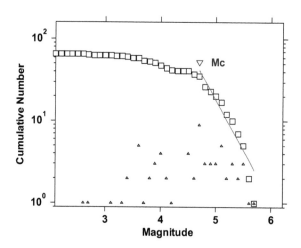

Maximum Likelihood Estimate, Uncertainties by bootstrapping
b-value = 1.22 +/- 0.61, a value = 7.37, a value (annual) = 5.73
Magnitude of Completeness = 4.7 +/- 0.36

Figure B.43 Frequency magnitude distribution for source zone 43

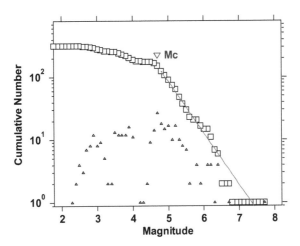

Maximum Likelihood Estimate, Uncertainties by bootstrapping
b-value = 0.83 +/- 0.1, a value = 6.07, a value (annual) = 3.75
Magnitude of Completeness = 4.7 +/- 0.16

Figure B.44 Frequency magnitude distribution for source zone 44

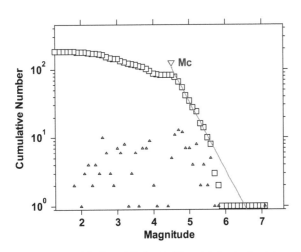

Maximum Likelihood Estimate, Uncertainties by bootstrapping
b-value = 1 +/- 0.36, a value = 6.54, a value (annual) = 4.24
Magnitude of Completeness = 4.5 +/- 0.69

Figure B.45 Frequency magnitude distribution for source zone 45

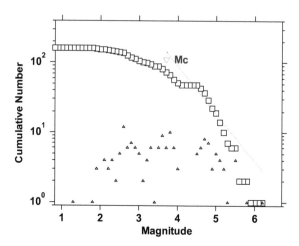

Maximum Likelihood Estimate, Uncertainties by bootstrapping
b-value = 0.69 +/- 0.39, a value = 4.7, a value (annual) = 2.8
Magnitude of Completeness = 3.7 +/- 0.87

Figure B.46 Frequency magnitude distribution for source zone 46

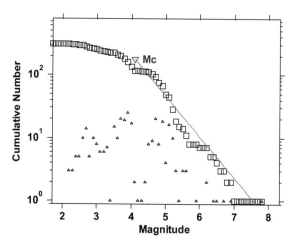

Maximum Likelihood Estimate, Uncertainties by bootstrapping
b-value = 0.65 +/- 0.25, a value = 4.9, a value (annual) = 2.64
Magnitude of Completeness = 4.1 +/- 0.44

Figure B.47 Frequency magnitude distribution for source zone 47

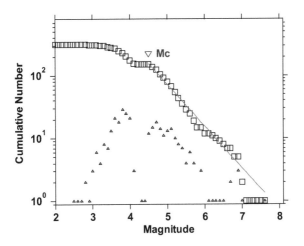

Maximum Likelihood Estimate, Uncertainties by bootstrapping
b-value = 0.67 +/- 0.14, a value = 5.18, a value (annual) = 2.52
Magnitude of Completeness = 4.5 +/- 0.34

Figure B.48 Frequency magnitude distribution for source zone 48

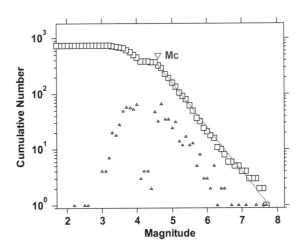

Maximum Likelihood Estimate, Uncertainties by bootstrapping
b-value = 0.8 +/- 0.07, a value = 6.19, a value (annual) = 3.96
Magnitude of Completeness = 4.6 +/- 0.14

Figure B.49 Frequency magnitude distribution for source zone 49

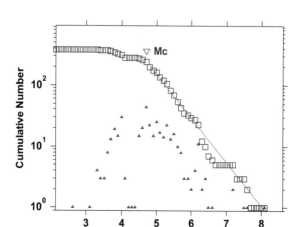

Maximum Likelihood Estimate, Uncertainties by bootstrapping
b-value = 0.72 +/- 0.09, a value = 5.76, a value (annual) = 3.55
Magnitude of Completeness = 4.7 +/- 0.19

Figure B.50 Frequency magnitude distribution for source zone 50

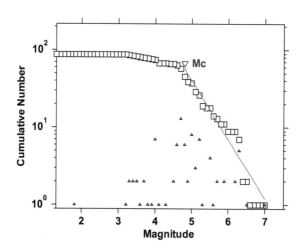

Maximum Likelihood Estimate, Uncertainties by bootstrapping
b-value = 0.76 +/- 0.33, a value = 5.41, a value (annual) = 3.03
Magnitude of Completeness = 4.8 +/- 0.24

Figure B.51 Frequency magnitude distribution for source zone 51

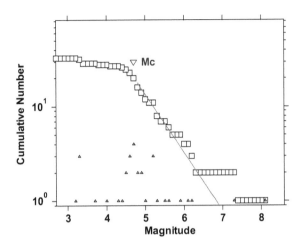

Maximum Likelihood Estimate, Uncertainties by bootstrapping
b-value = 0.62 +/- 0.14, a value = 4.21, a value (annual) = 2.07
Magnitude of Completeness = 4.7 +/- 0.58

Figure B.52 Frequency magnitude distribution for source zone 52

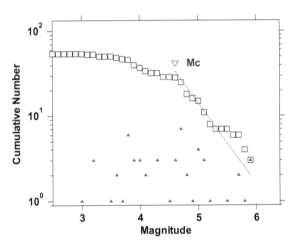

Maximum Likelihood Estimate, Uncertainties by bootstrapping
b-value = 0.95 +/- 0.43, a value = 5.89, a value (annual) = 3.64
Magnitude of Completeness = 4.6 +/- 0.49

Figure B.53 Frequency magnitude distribution for source zone 53

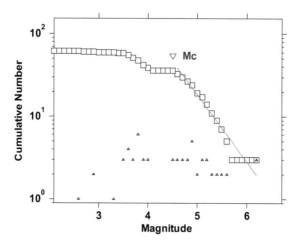

Maximum Likelihood Estimate, Uncertainties by bootstrapping
b-value = 0.81 +/- 0.34, a value = 5.33, a value (annual) = 3.41
Magnitude of Completeness = 4.5 +/- 0.71

Figure B.54 Frequency magnitude distribution for source zone 54

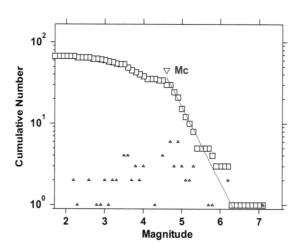

Maximum Likelihood Estimate, Uncertainties by bootstrapping
b-value = 0.93 +/- 0.42, a value = 5.84, a value (annual) = 3.74
Magnitude of Completeness = 4.6 +/- 0.57

Figure B.55 Frequency magnitude distribution for source zone 55

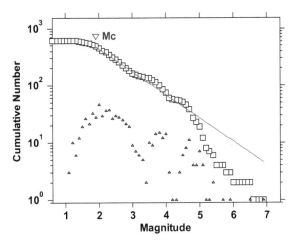

Maximum Likelihood Estimate, Uncertainties by bootstrapping
b-value = 0.4 +/- 0.02, a value = 3.45, a value (annual) = 1.95
Magnitude of Completeness = 1.9 +/- 0.14

Figure B.56 Frequency magnitude distribution for source zone 56

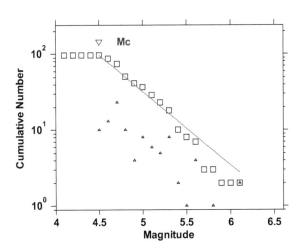

Maximum Likelihood Estimate, Uncertainties by bootstrapping
b-value = 0.97 +/- 0.08, a value = 6.33, a value (annual) = 4.79
Magnitude of Completeness = 4.5 +/- 0

Figure B.57 Frequency magnitude distribution for source zone 57

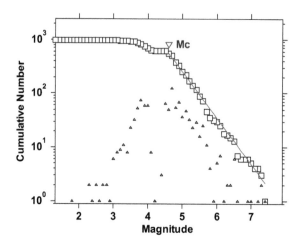

Maximum Likelihood Estimate, Uncertainties by bootstrapping
b-value = 0.86 +/- 0.08, a value = 6.7, a value (annual) = 4.17
Magnitude of Completeness = 4.6 +/- 0.09

Figure B.58 Frequency magnitude distribution for source zone 58

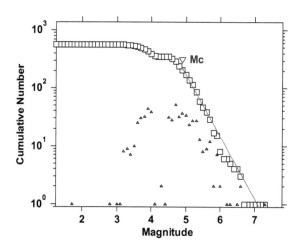

Maximum Likelihood Estimate, Uncertainties by bootstrapping
b-value = 1.07 +/- 0.19, a value = 7.57, a value (annual) = 4.86
Magnitude of Completeness = 4.9 +/- 0.25

Figure B.59 Frequency magnitude distribution for source zone 59

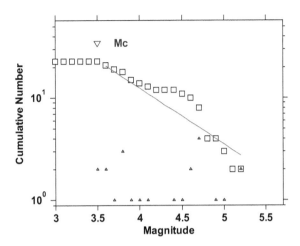

Maximum Likelihood Estimate, Uncertainties by bootstrapping
b-value = 0.55 +/- 0.08, a value = 3.28, a value (annual) = 1.67
Magnitude of Completeness = 3.5 +/- 0.04

Figure B.60 Frequency magnitude distribution for source zone 60

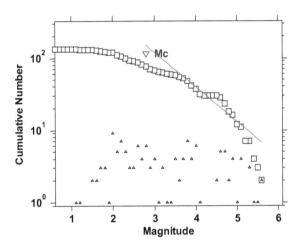

Maximum Likelihood Estimate, Uncertainties by bootstrapping
b-value = 0.48 +/- 0.33, a value = 3.53, a value (annual) = 1.64
Magnitude of Completeness = 2.8 +/- 0.98

Figure B.61 Frequency magnitude distribution for source zone 61

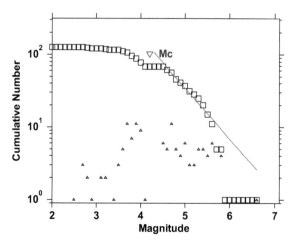

Maximum Likelihood Estimate, Uncertainties by bootstrapping
b-value = 0.7 +/- 0.47, a value = 5.05, a value (annual) = 3.04
Magnitude of Completeness = 4.2 +/- 0.5

Figure B.62 Frequency magnitude distribution for source zone 62

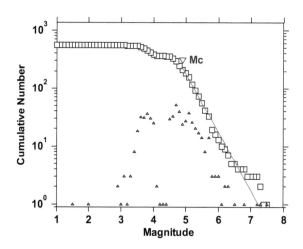

Maximum Likelihood Estimate, Uncertainties by bootstrapping
b-value = 1.01 +/- 0.17, a value = 7.29, a value (annual) = 5.12
Magnitude of Completeness = 4.9 +/- 0.2

Figure B.63 Frequency magnitude distribution for source zone 63

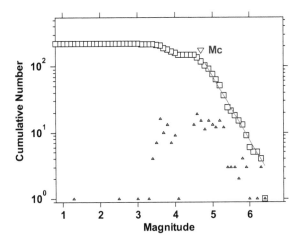

Maximum Likelihood Estimate, Uncertainties by bootstrapping
b-value = 0.97 +/- 0.27, a value = 6.72, a value (annual) = 4.78
Magnitude of Completeness = 4.7 +/- 0.41

Figure B.64 Frequency magnitude distribution for source zone 64

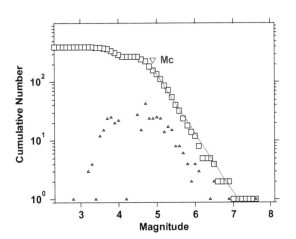

Maximum Likelihood Estimate, Uncertainties by bootstrapping
b-value = 1 +/- 0.17, a value = 7.15, a value (annual) = 4.61
Magnitude of Completeness = 4.9 +/- 0.21

Figure B.65 Frequency magnitude distribution for source zone 65

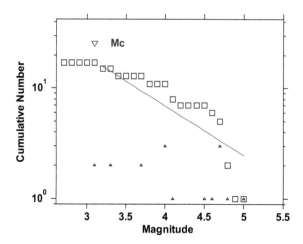

Maximum Likelihood Estimate, Uncertainties by bootstrapping
b-value = 0.45 +/- 0.08, a value = 2.63, a value (annual) = 0.959
Magnitude of Completeness = 3.1 +/- 0.08

Figure B.66 Frequency magnitude distribution for source zone 66

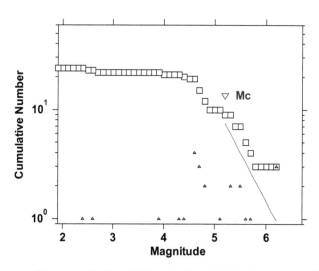

Maximum Likelihood Estimate, Uncertainties by bootstrapping
b-value = 0.89 +/- 0, a value = 5.49, a value (annual) = 3.19
Magnitude of Completeness = 5.2 +/- 0.7

Figure B.67 Frequency magnitude distribution for source zone 67

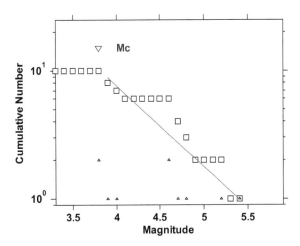

Maximum Likelihood Estimate, Uncertainties by bootstrapping
b-value = 0.63 +/- 0.17, a value = 3.42, a value (annual) = 1.22
Magnitude of Completeness = 3.8 +/- 0.04

Figure B.68 Frequency magnitude distribution for source zone 68

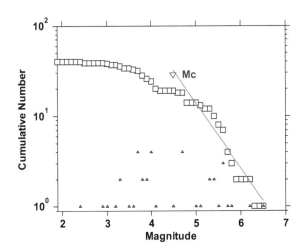

Maximum Likelihood Estimate, Uncertainties by bootstrapping
b-value = 0.72 +/- 0.46, a value = 4.77, a value (annual) = 2.54
Magnitude of Completeness = 4.5 +/- 0.68

Figure B.69 Frequency magnitude distribution for source zone 69

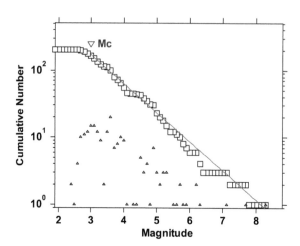

Maximum Likelihood Estimate, Uncertainties by bootstrapping
b-value = 0.43 +/- 0.08, a value = 3.53, a value (annual) = 1.25
Magnitude of Completeness = 3 +/- 0.43

Figure B.70 Frequency magnitude distribution for source zone 70

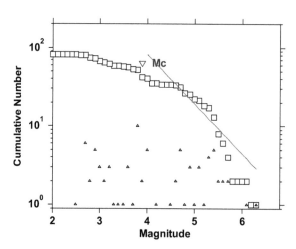

Maximum Likelihood Estimate, Uncertainties by bootstrapping
b-value = 0.63 +/- 0.5, a value = 4.44, a value (annual) = 2.28
Magnitude of Completeness = 3.9 +/- 0.69

Figure B.71 Frequency magnitude distribution for source zone 71

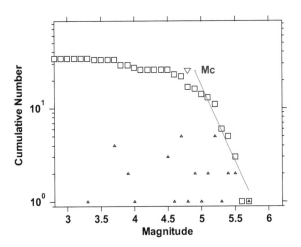

Maximum Likelihood Estimate, Uncertainties by bootstrapping
b-value = 1.62 +/- 0.69, a value = 9.33, a value (annual) = 7.43
Magnitude of Completeness = 4.8 +/- 0.53

Figure B.72 Frequency magnitude distribution for source zone 72

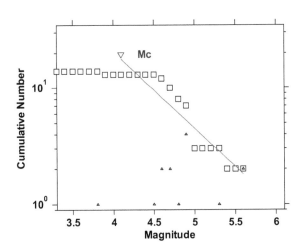

Maximum Likelihood Estimate, Uncertainties by bootstrapping
b-value = 0.66 +/- 0.33, a value = 3.95, a value (annual) = 2.3
Magnitude of Completeness = 4.1 +/- 0.35

Figure B.73 Frequency magnitude distribution for source zone 73

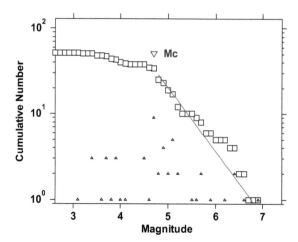

Maximum Likelihood Estimate, Uncertainties by bootstrapping
b-value = 0.74 +/- 0.14, a value = 5.02, a value (annual) = 2.64
Magnitude of Completeness = 4.7 +/- 0.25

Figure B.74 Frequency magnitude distribution for source zone 74

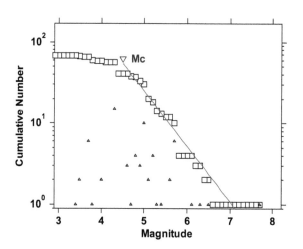

Maximum Likelihood Estimate, Uncertainties by bootstrapping
b-value = 0.68 +/- 0.22, a value = 4.8, a value (annual) = 2.37
Magnitude of Completeness = 4.5 +/- 0.34

Figure B.75 Frequency magnitude distribution for source zone 75

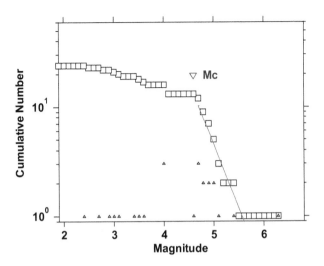

Maximum Likelihood Estimate, Uncertainties by bootstrapping
b-value = 1.17 +/- 0, a value = 6.5, a value (annual) = 4.81
Magnitude of Completeness = 4.6 +/- 0.56

Figure B.76 Frequency magnitude distribution for source zone 76

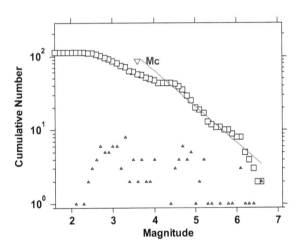

Maximum Likelihood Estimate, Uncertainties by bootstrapping
b-value = 0.49 +/- 0.3, a value = 3.75, a value (annual) = 1.54
Magnitude of Completeness = 3.6 +/- 0.81

Figure B.77 Frequency magnitude distribution for source zone 77

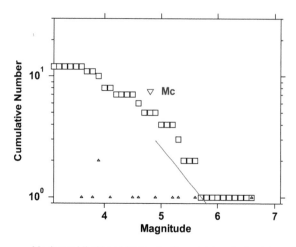

Maximum Likelihood Estimate, Uncertainties by bootstrapping
b-value = 0.58 +/- 0, a value = 3.29, a value (annual) = 1.39
Magnitude of Completeness = 4.8 +/- 0.91

Figure B.78 Frequency magnitude distribution for source zone 78

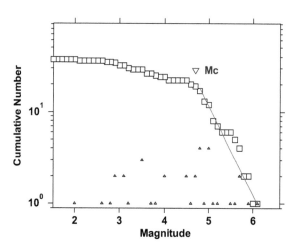

Maximum Likelihood Estimate, Uncertainties by bootstrapping
b-value = 0.97 +/- 0.15, a value = 5.9, a value (annual) = 3.76
Magnitude of Completeness = 4.7 +/- 0.66

Figure B.79 Frequency magnitude distribution for source zone 79

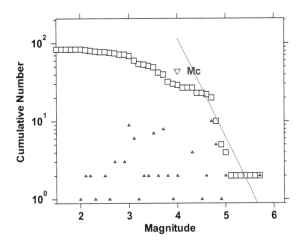

Maximum Likelihood Estimate, Uncertainties by bootstrapping
b-value = 1.27 +/- 0.88, a value = 7.15, a value (annual) = 4.89
Magnitude of Completeness = 4 +/- 0.72

Figure B.80 Frequency magnitude distribution for source zone 80

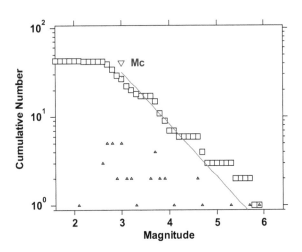

Maximum Likelihood Estimate, Uncertainties by bootstrapping
b-value = 0.59 +/- 0.19, a value = 3.26, a value (annual) = 1.12
Magnitude of Completeness = 3 +/- 0.37

Figure B.81 Frequency magnitude distribution for source zone 81

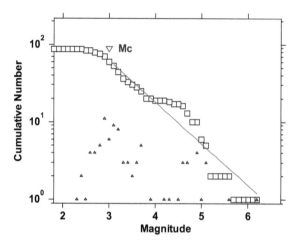

Maximum Likelihood Estimate, Uncertainties by bootstrapping
b-value = 0.54 +/- 0.07, a value = 3.4, a value (annual) = 0.426
Magnitude of Completeness = 3 +/- 0.19

Figure B.82 Frequency magnitude distribution for source zone 82

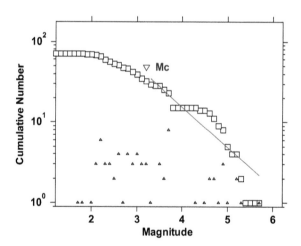

Maximum Likelihood Estimate, Uncertainties by bootstrapping
b-value = 0.5 +/- 0.12, a value = 3.19, a value (annual) = 0.97
Magnitude of Completeness = 3.2 +/- 0.62

Figure B.83 Frequency magnitude distribution for source zone 83

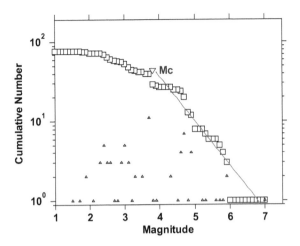

Maximum Likelihood Estimate, Uncertainties by bootstrapping
b-value = 0.56 +/- 0.21, a value = 3.81, a value (annual) = 1.19
Magnitude of Completeness = 3.8 +/- 0.48

Figure B.84 Frequency magnitude distribution for source zone 84

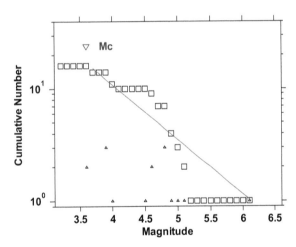

Maximum Likelihood Estimate, Uncertainties by bootstrapping
b-value = 0.49 +/- 0.09, a value = 2.99, a value (annual) = 1.82
Magnitude of Completeness = 3.6 +/- 0.11

Figure B.85 Frequency magnitude distribution for source zone 85

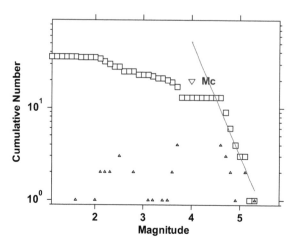

Maximum Likelihood Estimate, Uncertainties by bootstrapping
b-value = 1.24 +/- 0.75, a value = 6.67, a value (annual) = 4.27
Magnitude of Completeness = 4 +/- 0.91

Figure B.86 Frequency magnitude distribution for source zone 86

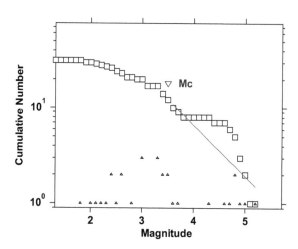

Maximum Likelihood Estimate, Uncertainties by bootstrapping
b-value = 0.54 +/- 0, a value = 2.97, a value (annual) = 0.806
Magnitude of Completeness = 3.5 +/- 0.82

Figure B.87 Frequency magnitude distribution for source zone 87

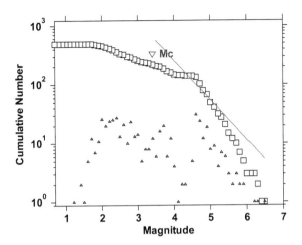

Maximum Likelihood Estimate, Uncertainties by bootstrapping
b-value = 0.67 +/- 0.38, a value = 5.1, a value (annual) = 2.71
Magnitude of Completeness = 3.4 +/- 1.14

Figure B.88 Frequency magnitude distribution for source zone 88

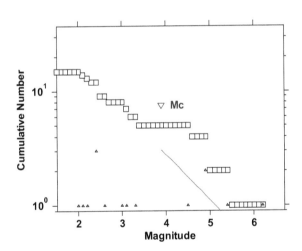

Maximum Likelihood Estimate, Uncertainties by bootstrapping
b-value = 0.4 +/- 0, a value = 2.05, a value (annual) = 0.297
Magnitude of Completeness = 3.9 +/- 1.41

Figure B.89 Frequency magnitude distribution for source zone 89

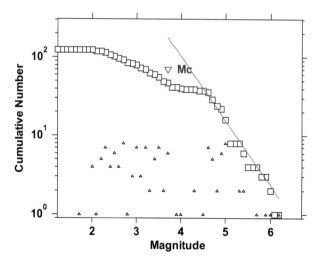

Maximum Likelihood Estimate, Uncertainties by bootstrapping
b-value = 0.82 +/- 0.64, a value = 5.29, a value (annual) = 2.97
Magnitude of Completeness = 3.7 +/- 1.04

Figure B.90 Frequency magnitude distribution for source zone 90

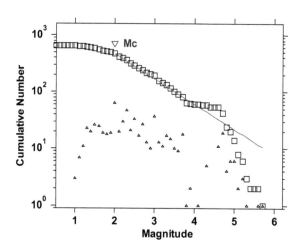

Maximum Likelihood Estimate, Uncertainties by bootstrapping
b-value = 0.44 +/- 0.02, a value = 3.56, a value (annual) = 1.29
Magnitude of Completeness = 2 +/- 0.08

Figure B.91 Frequency magnitude distribution for source zone 91

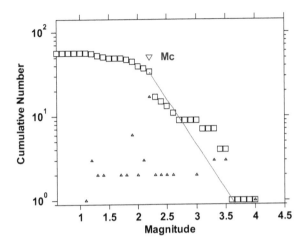

Maximum Likelihood Estimate, Uncertainties by bootstrapping
b-value = 1.08 +/- 0.22, a value = 3.89, a value (annual) = 2.37
Magnitude of Completeness = 2.2 +/- 0.04

Figure B.92 Frequency magnitude distribution for source zone 92

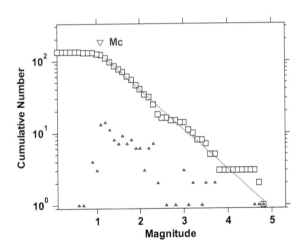

Maximum Likelihood Estimate, Uncertainties by bootstrapping
b-value = 0.55 +/- 0.07, a value = 2.7, a value (annual) = 0.794
Magnitude of Completeness = 1.1 +/- 0.17

Figure B.93 Frequency magnitude distribution for source zone 93

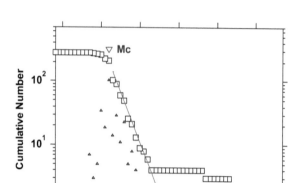

Maximum Likelihood Estimate, Uncertainties by bootstrapping
b-value = 1.58 +/- 0.19, a value = 5.8, a value (annual) = 3.62
Magnitude of Completeness = 2.2 +/- 0.03

Figure B.94 Frequency magnitude distribution for source zone 94

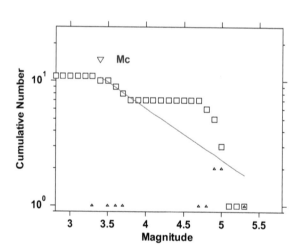

Maximum Likelihood Estimate, Uncertainties by bootstrapping
b-value = 0.41 +/- 0.11, a value = 2.44, a value (annual) = 0.193
Magnitude of Completeness = 3.4 +/- 0.11

Figure B.95 Frequency magnitude distribution for source zone 96

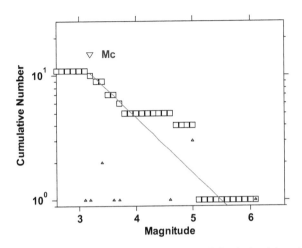

Maximum Likelihood Estimate, Uncertainties by bootstrapping
b-value = 0.45 +/- 0.13, a value = 2.44, a value (annual) = 1.03
Magnitude of Completeness = 3.2 +/- 0.1

Figure B.96 Frequency magnitude distribution for source zone 97

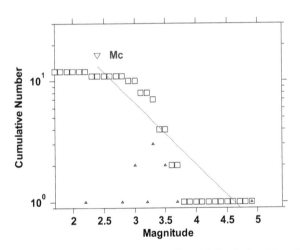

Maximum Likelihood Estimate, Uncertainties by bootstrapping
b-value = 0.51 +/- 0.22, a value = 2.34, a value (annual) = 0.765
Magnitude of Completeness = 2.4 +/- 0.3

Figure B.97 Frequency magnitude distribution for source zone 100

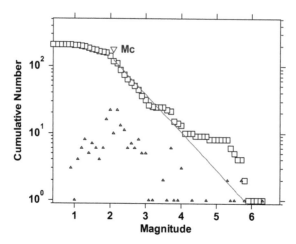

Maximum Likelihood Estimate, Uncertainties by bootstrapping
b-value = 0.57 +/- 0.06, a value = 3.28, a value (annual) = 1.03
Magnitude of Completeness = 2.1 +/- 0.14

Figure B.98 Frequency magnitude distribution for source zone 101

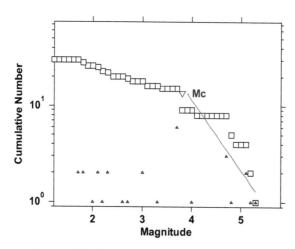

Maximum Likelihood Estimate, Uncertainties by bootstrapping
b-value = 0.73 +/- 0.23, a value = 3.97, a value (annual) = 1.72
Magnitude of Completeness = 3.8 +/- 0.56

Figure B.99 Frequency magnitude distribution for source zone 102

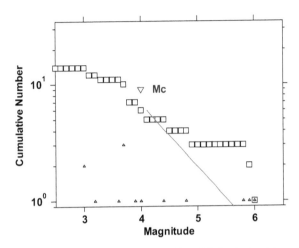

Maximum Likelihood Estimate, Uncertainties by bootstrapping
b-value = 0.54 +/- 0, a value = 2.99, a value (annual) = 0.72
Magnitude of Completeness = 4 +/- 0.94

Figure B.100 Frequency magnitude distribution for source zone 103

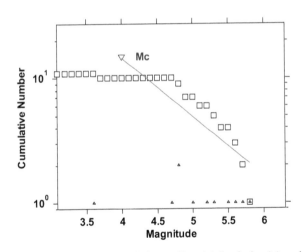

Maximum Likelihood Estimate, Uncertainties by bootstrapping
b-value = 0.47 +/- 0.26, a value = 3.05, a value (annual) = 1.28
Magnitude of Completeness = 4 +/- 0.54

Figure B.101 Frequency magnitude distribution for source zone 104

Appendix C: Results of PSHA with equal weighting scheme

The probabilistic seismic hazard analysis as described in Chapter 5 was performed with different combinations of weights assigned to various models (for source model, M_{max} estimation, and ground motion prediction equations). The results obtained by adopting an equal weighting scheme for all the models are presented here.

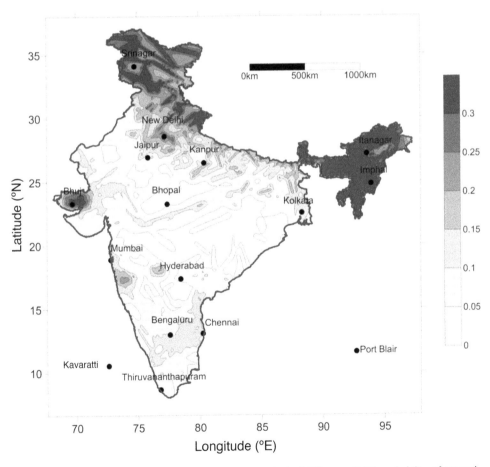

Figure C.1 PGA values (g) corresponding to a return period of 475 years (10% probability of exceedance in 50 years)

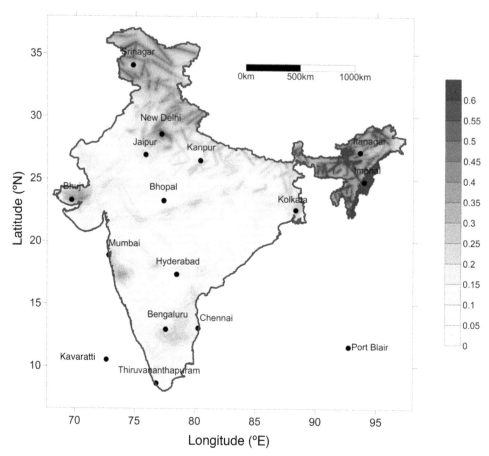

Figure C.2 PGA values (g) corresponding to a return period of 2475 years (2% probability of exceedance in 50 years

The spatial variations of the PGA value corresponding to 10% and 2% probability of exceedance in 50 years are shown in Figures C.1 and C.2, respectively. The PGA values obtained from PSHA using the logic tree with different weights (Chapter 6) were compared with those obtained with equal weights (Table C.1). The PGA values obtained from the former analysis are on higher side.

Table C.1 PGA values at rock level for the 10 most populous cities of India

Major cities	Location		PGA value (g)			
			Logic tree with varied weights		Logic tree with equal weights	
	Longitude (°E)	Latitude (°N)	For return period of 475 years	For return period of 2475 years	For return period of 475 years	For return period of 2475 years
Mumbai	72.82	18.90	0.10	0.19	0.10	0.18
Delhi	77.20	28.58	0.27	0.51	0.24	0.45
Bangalore	77.59	12.98	0.13	0.22	0.11	0.21
Kolkata	88.33	22.53	0.13	0.23	0.13	0.22
Chennai	80.25	13.07	0.13	0.21	0.10	0.18
Hyderabad	78.48	17.38	0.01	0.19	0.08	0.15
Ahmedabad	72.62	23.00	0.10	0.18	0.10	0.16
Pune	73.87	18.53	0.07	0.13	0.07	0.12
Kanpur	80.40	26.47	0.11	0.17	0.11	0.20
Jaipur	75.87	26.92	0.12	0.2	0.10	0.17

Appendix D: Preparation of the slope map

D.1 INTRODUCTION

This appendix presents the report on the development of slope map from satellite images, as described in Chapter 6.

D.1.1 Input details

Input data: Elevation data for the area of interest from Cartosat DEM, SRTM, Resampled SRTM, Russian 1:200,000 topo maps.

Input data source: National Remote Sensing Centre Bhuvan, USGS, and Consortium for Spatial Information of the Consultative Group for International Agricultural Research (CGIAR CSI), and Viewfinder Panaromas.

Operating platform: GRASS 6.4 RC1 on Fedora 14.

D.1.2 Output details

Output: Slope map on a $0.1° \times 0.1°$ cell size.

Area of interest (Ai): The whole of the political boundary of India with the rectangular bounds given below:

N: 38N S: 8N
E: 98E W: 68E

D.2 PROCESS DETAILS

Steps involved in the generation of the slope map for India:

D.2.1 Data download

The data were downloaded from various sources, including the Cartosat Digital Elevation Model (DEM) from NRSC Bhuvan, SRTM from USGS Earth Resources Observation and Science (USGS EROS), interpolated seamless SRTM data v4.1 from CGIAR CSI, and void-filled data in mountainous terrain from viewfinder panoramas.

Details of downloaded data:

a.   SRTM DATA:
    i.   UTM grid coverage: 28 grids
    ii. Total number of data: ~ 390 cells
    iii. Area of land covered in each data file: $1° \times 1°$ cell
    iv. Resolution: 3" (90 m)

b.   Cartosat DEM:
    i.   UTM grid coverage: 28 grids
    ii. Total number of data: ~ 390 cells
    iii. Area of land covered in each data file: $1° \times 1°$ cell
    iv. Resolution: 1" (30 m)

c.   Seamless data from CIGAR CSI:
    i.   UTM grid coverage: 28 grids
    ii. Total number of data: ~ 15 cells
    iii. Area of land covered in each data file: $5° \times 5°$ cell
    iv. Resolution: 3" (90 m)

d.   Mosaic from viewfinder panaromas:
    i.   UTM grid coverage: 28 grids
    ii. Total number of data: ~ 390 cells
    iii. Area of land covered in each data file: $1° \times 1°$ cell
    iv. Resolution: 3" (90 m)

D.2.2   Georeferencing

The available input data available were georeferenced based on the WGS84 datum and lat/long coordinate system on the UTM projection system. Hence, the data were readily available for further processing and mosaicking.

D.2.3   Resampling

All the available data were resampled to a common map resolution of 6′ (0.1°) using the nearest neighborhood method. However, areas with large undulation and steep slopes such as the Himalayan regions were resampled using spline interpolation method to maintain the terrain slope.

D.2.4   Mosaicking

Because the original data were already georeferenced, mosaicking was readily carried out on the resampled data. The result was data for the entire area of interest within the political boundary of India on a map scale of 0.1° resolution.

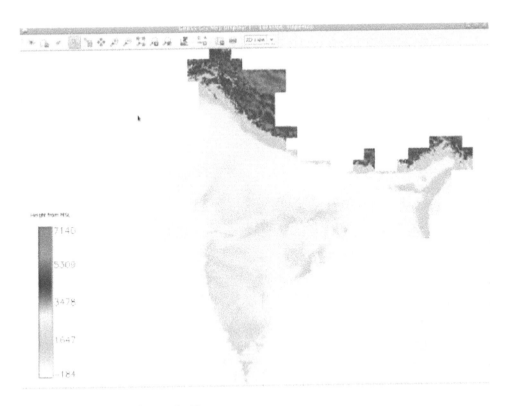

Figure D.1 Resampled and mosaicked image

D.2.5   Error correction

Because the main data used was that of elevation values, obtained from SRTM mission by NASA in February 2000, the main errors were outliers and NULL data. Where outliers were caused by sources like open cut mines, multiple reflection, or reflection from near vertical surfaces, they were characterized by negative or NULL values. However, other predominant sources of NULL data were the Himalayan region and steep hilly regions. These were the result of no data being recorded for the steeply sloping region of the hills that lies on the other side of the satellite path during the scanning by the side-looking radars. These NULL regions were filled with the help of data obtained from secondary sources like the Cartosat DEM and Russian 1:200,000 topographic maps, among others.

a.   Removal of negative and spurious data

After mosaicking, the negative data were removed by using a threshold filter with a cutoff value of 0 m and 9000 m to eliminate all data that were beyond the cutoff values and convert them to NULL values, as the highest point within the Indian region has an elevation of 8848 m.

Metadata of a region containing negative values (note the minimum value in the range of data):

Type of map: Raster
Number of Categories: 2482
Data type: CELL|
Rows: 40
Columns: 270
Total cells: 10,800
Projection: Latitude-longitude
N: 24N S: 20N Res: 0:06
E: 95E W: 68E Res: 0:06
Range of data: Min = −7 max = 2482

b.   Filling up of NULL regions

The NULL regions were filled with the help of secondary data, using the map calculator function of GRASS 6.4 RC1. The voids (NULL) were removed and replaced with interpolated values using the spline fitting method to accommodate surface undulation in mountainous terrain.

D.3   GENERATION OF THE SLOPE MAP FOR THE ENTIRE AREA OF INTEREST FROM THE ABOVE DATA

The slope map was generated using the slope map function of GRASS 6.4 RC1. The algorithm used for the generation of the slope map by GRASS is given by the Grass Manual as follows:

The algorithm used to determine slope and aspect uses a 3 × 3 neighborhood around each cell in the elevation file. Thus, it is not possible to determine slope and aspect of the cells adjacent to the edges in the elevation map layer. These cells are assigned a "zero slope" value (category 0) in both the slope and aspect raster map layers.

Horn's formula is used to find the derivatives in x and y directions.

Only when using integer elevation models, the aspect is biased in 0, 45, 90, 180, 225, 270, 315, and 360 directions; i.e., the distribution of aspect categories is very uneven, with peaks at 0, 45, . . . , 360 categories. Because most cells with a minimal slope end up having category 0, 45, . . . , 360 it is sometimes possible to reduce bias in these directions by filtering out the computation of aspect in areas where the terrain is almost flat. The new option min_slp = value was added (minimum slope for which aspect is computed). The aspect for all cells with slope < min_slp is set to 0 (no value). When working with floating point elevation models, no such aspect bias occurs.

Shown below is the metadata for the original map and the slope map, which shows the reduced boundary size because slope could not be derived for the edge regions as per the above algorithm.

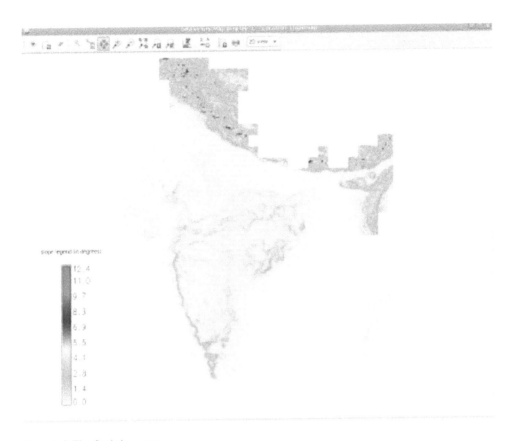

Figure D.2 The final slope map

1. Initial rectangular regional boundary enclosing the area of interest:

    _

    Type of map: Raster
    Number of categories: 255
    Data type: FCELL
    Rows: 300
    Columns: 300
    Total cells: 90,000
    Projection: Latitude-longitude
    N: 38N S: 8N Res: 0:06
    E: 98E W: 68E Res: 0:06
    Range of data: min = 0 max = 6664

2. An approximate effective region of the slope map:

    _ _ _ _ _ _ _ _ _ _ _ _ _ _ _ _ _ _

    Type of map: Raster
    Number of Categories: 255

Data type: FCELL
Rows: 304
Columns: 294
Total cells: 89,376
Projection: Latitude-longitude
N: 36:57:00N S: 8:07:00N Res: 0:06
E: 97:25:40E W: 68:15:17E Res: 0:06
Range of data: min = 0 max = 12.42135
zfactor = 1.00
Format = Degrees
min_slp_allowed = 0.000000

REFERENCE

1. Horn, B.K.P. (1981) Hill Shading and the Reflectance Map. *Proceedings of the IEEE*, 69(1), 14–47.

Data Sources

1. www.viewfinderpanoramas.org/ dem3/DEM3-ReadMe.html.
2. www.cgiar-csi.org/data/elevation/item/45-srtm-90 m-digital-elevation-database-v41.
3. http://bhuvan-noeda.nrsc.gov.in/download/download/download.php.
4. http://eros.usgs.gov/#/Find_Data/Products_and_Data_Available/SRTM.

Index

Note: Page numbers in italics indicate figures. Page numbers in bold indicate tables.

aftershocks 22–25, **23–24**, *24*, **25**, *25*
areal sources 34
attenuation models 37–39, **38**
attenuation relation *see* ground motion
    prediction equation (GMPE)
a values 46–51, *51*

body wave magnitude 2–3
b values 46–51, *49*, **50**, *50*

completeness analysis 25–26
crustal region, shallow 8

database, earthquake, compilation of 17–18, *18*
delineation of seismic source zones 51–52, *52*
deterministic seismic hazard assessment
    (DSHA): attenuation models in 37–39, **38**;
    hazard estimation in 36–39, **38**; logic tree
    structure in 40, *40*; methodology in 35, *36*;
    peak ground acceleration in *87*
DSHA *see* deterministic seismic hazard
    assessment (DSHA)

earthquake: defined 1
earthquake database, compilation of 17–18, *18*
earthquake size 1–3
epicenter 1, *2*
Eurocode-8 **81**, 81–83, **82**

fault place *2*
foreshocks 22–25, **23–24**, *24*, **25**, *25*
frequency magnitude distribution (FMD): in
    magnitude of completeness 46; plots *113–163*;
    in seismicity analysis 45; in seismicity
    parameters 54

georeferencing 30, *30*
GMPE *see* ground motion prediction equation
    (GMPE)
gridded seismicity source model 33–34
ground motion evaluation: in probabilistic
    seismic hazard assessment
    61–64, *63*
ground motion prediction equation (GMPE)
    9–12, **10**, *11*, 37–39, **38**

homogenization, of earthquake magnitudes
    19–22, *20–21*, **22**
hypocenter 1, *2*
hypocentral uncertainty evaluation
    in 63, *63*

India, geological setting of 3–5, *4*
Indian subcontinent: probabilistic seismic hazard
    assessment case study with 64–75, *65–66*, **67**,
    67, *69–73*, **74–75**
Indian subcontinent, seismotectonics 5–9,
    6–7
intensity 1–2

linear seismic sources 31–32
liquefaction: defined 78; geophysical data
    and 80; geotechnical data and 79, **80**;
    peak ground acceleration and **86**, 86–89,
    87–90, **90**; potential evaluation 91–95,
    93, *94–95*; site classification and 78–83,
    79–82; site classification schemes in **81**,
    81–83, **82**; surface geology and 78–79,
    79; V$_S$^{30} map and 83–84, *84*, **85**,
    85, **86**
local site effect 77, *77*

magnitude: homogenization of 19–22, *20–21*, **22**;
    maximum possible 52–53; overview of 2–3
magnitude conversion relations *109*, **110**, *110*, **111**
magnitude of completeness 46
magnitude recurrence rate 61–63
main shock identification 22–25, **23–24**, *24*, **25**, *25*
map, seismotectonic 26–30, **27–29**, *29–30*
Medvedev-Sponheuer-Karnik (MSK) scale 2
Mercalli-Cancani-Selberg (MCS) scale 2
Modified Mercalli (MM) scale 2
moment magnitude 3, 19–22, *20–21*, **22**

National Earthquake Hazard Reduction Program
    (NEHRP) 78–79, **81**, 81–83, **82**
NEHRP *see* National Earthquake Hazard
    Reduction Program (NEHRP)

past efforts, of seismic hazard studies 13–16, **14**,
    15–16
peak ground acceleration (PGA) 9, *11*, 13, *15*, 35,
    40, **41**, *41–43*, 43–44, *44*, 59, 65–75, *66*, **67**,
    67, *69–73*, **74–75**, 77, **86**, 86–89, *87–90*, **90**
PGA *see* peak ground acceleration (PGA)
point sources 32–33
probabilistic seismic hazard assessment (PSHA):
    attenuation relationship uncertainty in 64;
    case study in, with Indian subcontinent
    64–75, *65–66*, **67**, *67*, *69–73*, **74–75**; ground
    motion evaluation in 61–64, *63*; hypocentral
    uncertainty evaluation in 63, *63*; in liquefaction
    potential evaluation 92; logic tree structure
    in 64–65, *65*; magnitude recurrence rate in
    61–63; methodology 59–60, *60*; peak ground
    acceleration in 65–75, *66*, **67**, *67*, *69–73*,
    74–75, 87–88, *88*; results, with equal weighting
    scheme *165*, **166**, *166*; steps in 60, *60*
PSHA *see* probabilistic seismic hazard
    assessment (PSHA)

seismic gap 12–13, *13*
seismic hazard assessment *see* deterministic
    seismic hazard assessment (DSHA);
    probabilistic seismic hazard assessment
    (PSHA)
seismic hazard studies, past efforts of 13–16, **14**,
    15–16
seismicity analysis 45–51, *47*, *49*, **50**,
    50–51
seismicity parameters, for source zones 54, *54*,
    55–57
seismic source models 31–34
seismic source zones, delineation of 51–52, *52*
seismic zonation 7
seismotectonic map 26–30, **27–29**, *29–30*
seismotectonics, of Indian subcontinent 5–9,
    6–7
shallow crustal region 8
site classification: geophysical data in 80;
    geotechnical data in 79, **80**; methods
    78–83, **79–82**; overview of 78; peak ground
    acceleration and **86**, 86–89, *87–90*, **90**;
    schemes **81**, 81–83, **82**; surface geology
    in 78–79, *79*; V_s^{30} map in 83–84, *84*, **85**,
    85, **86**
size, earthquake 1–3
slope map 83–84, *84*, **85**, *85*, **86**, 169–174,
    171, *173*
source models 31–34
source zones, estimation of seismicity
    parameters for 54, *54*, **55–57**
subduction zones 8–9
surface geology, liquefaction and 78–79, *79*
surface wave magnitude 2–3

tectonic provinces 7

V_s^{30} map 83–84, *84*, **85**, *85*, **86**